THE
MACMILLAN
BOOK
OF BERRY GARDENING

GÜNTHER LIEBSTER

THE MACMILLAN BOOK

OF BERRY GARDENING

Translated by Carole Ottesen

COLLIER BOOKS
MACMILLAN PUBLISHING COMPANY
New York

COLLIER MACMILLAN PUBLISHERS
London

Copyright © 1986 by
Macmillan Publishing Company,
a division of Macmillan, Inc.

Macmillan Publishing Company
866 Third Avenue, New York, N.Y. 10022
Collier Macmillan Canada, Inc.

Title of the original German edition:
BEERENOBST FÜR JEDEN GARTEN
© 1984 BLV Verlagsgesellschaft mbH,
München

Library of Congress Cataloging-in-Publication Data
Liebster, Günther.
  The Macmillan book of berry gardening.

  "Collier books."
  1. Berries   I. Title.   II. Title: Book of berry gardening.
SB381.L5413 1986   634.7   85-23209

ISBN 0-02-063360-2

Macmillan books are available at special
discounts for bulk purchases for sales
promotions, premiums, fund-raising, or
educational use.

For details, contact:

Special Sales Director
Macmillan Publishing Company
866 Third Avenue
New York, N.Y. 10022

10 9 8 7 6 5 4 3 2 1

Printed in Germany

Contents

Introduction

While this book, like the others in the series, was written with the home gardener and do-it-yourselfer in mind, commercial growers who wish to experiment with different varieties of berries or cultivation methods also may find it useful. On the other hand, the same home gardener, with no farm machinery at his disposal, no hourly labor quotient, and no need to keep accounts on crops, nonetheless can utilize it as an efficient means for learning the same techniques employed in commercial berry production (and, indeed, for him working in the garden will be a pleasant and refreshing hobby). Either way, delicious berries will be the result, and that's what this book is all about.

There is no good reason not to have a garden with berries. The kinds of berries included here grow on small bushes or plants that need very little room, and they bear fruit soon after planting and provide many years of harvests.

At first the crops will provide only a few luxurious delicacies—but very soon they will be harvested in such quantity that they can actually make a substantial dent in your shopping bill. Making marmalades, jellies, jams, compotes, drinks, and other delicious products from your own fruit not only saves money, but also provides important nutrients in the diet.

Certainly berries are found in the supermarket and shops, but in such small quantities and at such exorbitant prices that many people are scared away. When you consider the hourly wages involved in picking a berry crop, the prices may be fair, but they certainly don't encourage greater consumption of berries, and a large family with a small income may find that buying berries is a luxury beyond their means. All the more reason to grow them in your own garden, where the work of caring and harvesting can be done at leisure.

This book covers not only the "classical" berries—currants, gooseberries, raspberries, blackberries, and strawberries—but also the fruit-bearing species of the genus *Vaccinium*—blueberries, cowberries, cranberries. Many readers have had more experience eating these species than growing them—especially the cranberries and cowberries. Besides being a

The strawberry Rimona-Hummi.

Introduction

tradition at holiday tables, these berries provide nutritionally excellent sauces and beverages, and if the climate permits they are certainly worth trying. All members of the genus *Vaccinium* included here require highly acidic soil.

There are three new horticultural developments that may not yet have been written about: the jostaberry, a cross between currants and gooseberries, the new "strawberry meadow," and recent experiments with the spindle or pillar method of growing currants and gooseberries. Three chapters of this book explore these new developments.

From a botanical viewpoint, the elderberries (*Sambucus nigra*), the sea-buckthorns (*Hippophae rhamnoides*), the rose hips (*Rosa* sp.), the blackthorns (*Prunus spinosa*), the barberries (*Berberis* sp.), and the European cranberries (*Vaccinium oxycoccus*) should be included among the berries. Although these species may bear healthy and versatile fruits, they will not be covered here as the book's format of necessity limits material to be included. For each

An abundant harvest from the garden.

type of berry, only the most basic and useful cultivation practices, the most important varieties, and the most dangerous diseases and pests are discussed; further information must be sought in more specialized literature.

The objective of this book is to promote the cultivation of ever more bountiful berries—from the early summer–ripening types to those harvested throughout the summer into fall. The ravenous appetites and the enthusiasm with which our children (and not only children) devour berries are clear indications that we are growing far too few of them.

Berries Are Nutritious

The high nutritional value of berries has never been doubted. Not only do they contain astoundingly high amounts of different vitamins — especially vitamin C—they also contain minerals, sugars, and acids important for good health. Regular intake of these substances builds strong bodies capable of withstanding colds and other infectious diseases.

The following chart shows the nutritional elements contained in three and a half ounces of each fruit. Of course, the amounts listed are averages. Where variables of variety, location, weather, fertilization, and method are concerned, no hard and fast rules are possible. Yet no matter what the exact numbers, the range of vitamins and minerals contained in berries would be reason enough to grow them without even considering their excellent flavor.

Especially today, when we are actively seeking better natural foods, the extraordinarily high nutritional value of berries is worth considering.

Introduction

Nutritional Contents of Berries

		Straw-berries	Currants Red	Currants Black	Goose-berries	Rasp-berries	Black-berries	Blue-berries	Cow-berries	Cran-berries
Water	g	89–91	84.7	81.3	87.3	84.5	84.7	82.4	87.4	88.0
Carbohydrates	g	6–9	9.7	12.4	8.8	8.1	8.6	12.1.–15.1	9.9	4.2
Protein	g	0.6–1	1.1	1.3	0.8	1.3	1.2	0.6–0.7	0.3	0.2
Calories		55	45	57	44	40	48	61	46	26
Sodium	mg	1	1.4	1.5	1.6	4	3	1	2	2
Potassium	mg	145	238	310	203	170	189	65	72	53
Magnesium	mg	12	13	17	15	30	26	2.4	5.5	5.5
Calcium	mg	28	29	46	29	40	29	10	14	1.3
Iron	mg	1	0.9	1.3	0.6	1	0.9	0.8	0.5	0.4
Phosphorus	mg	25	27	40	30	44	30	13	9.7	8
Vitamin A	I.U.	60	120	120	290	130	200	280	65	40
Vitamin B_1	mg	0.05	0.04	0.05	0.02	0.02	0.03	0.02	0.01	0.01
Vitamin B_2	mg	0.05	0.03	0.04	0.02	0.05	0.04	0.02	0.02	
Vitamin B_6	mg		0.05	0.08		0.90	0.05	0.60	0.01	0.01
Vitamin C	mg	50–100	36	177	35	25	17	16	12	7.5–10.5
Total Sugar	g	3–10	5.61	6.99	6.03	5.58	11.31	11.45	8.73	4.20
Total Acid	g	0.7–3.0 C	2.38 C	3.29 C	1.60 T	1.40 T	0.7–4.2 C	0.71 C	2.40 C	2.40 C

I.U. = International Units; C = Citric acid; T = Tartaric acid

Introduction

One of the greatest pleasures for children is harvesting directly into the mouth.

Currants

Red and White Currants

Currants were first cultivated in the cloisters of medieval Europe; today, they are frequently found growing both commercially and in the home garden. White currants (actually currants are yellow) are not grown as widely commercially because they generally produce smaller harvests than red currants, but gardeners who are familiar with their excellent taste prize them nonetheless.

Botany

The species of currants are members of the *Ribes* genus of the Saxifragaceae family:
Ribes rubrum, from which most species are descended.
Ribes petraeum, a European species called "rock currant" that comes from the high mountains and is especially suited to rough terrain (Rote Hollaendische, Vierlaender, Weisse aus Jueterbog).
Ribes spicatum, the northern currant (Houghton Castle).
Ribes multiflorum (R. multiflorum × Rote Hollaendische = Rote Spaetlese).

The botanical name for all red and white species is generally *Ribes rubrum.*

Currants are mainly self-fruitful, but it is still beneficial to plant several varieties. Cross-pollination results in larger harvests and bigger berries. A beehive in the vicinity of the berries enhances fertilization and ensures the development of fruit.

Varieties

Red Currants

Jonkheer van Tets.   (1941). Early blooms and fruit from the end of June. A strong, upright but only lightly branching growth, it requires a setting out of the wind and protected from late frosts. It produces long clusters of large, thin-skinned, soft, dark red single berries. The taste is somewhat sour and aromatic, and it belongs to the finest-tasting species.

Heros.   A German hybrid (Hermann Rosenthal, 1927). It blooms early and ripens early to mid-season. Growth is weak, somewhat drooping, with little branching. Hard pruning and good fertilization are necessary. It produces long clusters with very large, red, mildly aromatic, bittersweet berries that are outstanding for eating fresh. Sometimes inferior specimens are sold.

Rote Vierlaender.   (Syn. Erstling aus Vierlanden). An English hybrid that blooms in mid-season with robust spreading growth. It prefers

Fruit clusters of the red currant species Heros.

a fertile soil. Blooms late; clusters are average to long, growing close together. Better suited for cooking and canning than for eating fresh. Harvests can yield up to 25 pounds per bush. Potassium deficiency shows up in necrosis on the edges of leaves.

Large cluster of the currant species Rote Vierlaender.

Currants

Flowers of Heinemanns Rote Spaetlese.

Heinemanns Rote Spaetlese.

Red Lake.   An American hybrid (1933) that is early. Growth is moderate; habit, broadly round. Clusters are long, packed with large, firm, medium red, mildly sour, aromatic fruit. High vitamin C content, good for eating fresh.

Rondom.   This Dutch hybrid (1949) is late. It requires good care and an advantageous situation. Growth is strong, upright, drooping when carrying fruit. Clusters are average, with berries growing thickly all around the branch (Rondom). The berries that are good for juice are firm, bright red, sour, and of high quality.

Rotet.   A new hybrid from Holland, late bearing. Growth is strong and spreading; large, dark, delicious berries in small harvests.

Heinemanns Rote Spaetlese.
A German hybrid (O. Macherauch, 1942), extremely late, bearing toward the end of July and into August. Growth average to strong, with unusually long clusters hanging thickly on the branches. Large harvests of medium-sized, light red, very firm, somewhat sour, scentless berries having many seeds. Good variety for fresh eating and preserving.

White Currants

<u>Weisse Versailler.</u>   French hybrid, known since 1850, ripening early on strong, upright plants. Long clusters of medium-sized, mildly sweet, aromatic berries are good for eating fresh.

<u>Weisse aus Jueterbog.</u>   Of unknown origin, this is an early ripening, weakly growing plant needing little room. Responds to fertilizing. Clusters are medium-sized; the berries average and deliciously aromatic. Produces only small harvests for fresh eating.

The white currants Weisse Versailler (*right*) and Weisse aus Jueterbog (*below*).

Currants

Propagation

Most gardeners buy young nursery plants, but growing one's own is not particularly difficult.

Necessary for propagation: Cuttings from robust growth from the current year about 6–8 inches long; one branch will yield two or three cuttings. Cut just below a bud at the base, making a slanting cut to slightly above a bud at the top. Space about 6 inches apart in light, well-drained soil in late September or October. Insert cuttings deeply so that only one or two buds are showing above the surface; in the fall, after they have developed calluses, roots will begin to grow. The second fall, the plants can be transplanted further apart. Cut back hard: The one or two branches can be cut back to three or four buds. The third fall, the plants are ready for the garden.

As with plants of all kinds, meristem culture has produced numerous currant varieties, and the rootstock *Ribes aureum* (the yellow currant) is abundant and easily available for the propagation of healthy new plants. Fruit-bearing stems of currants and gooseberries are grafted onto *Ribes aureum* rootstocks in lengths of 30–36 inches for tall plants, and 15–25 inches for low-growing ones. Usually the stem is a lateral branch with three well-developed shoots if a currant, and four if a gooseberry is used.

Cuttings of currants (*a, b*).
Cutting back after one year (*c*).
Currant ready to plant (*d*).

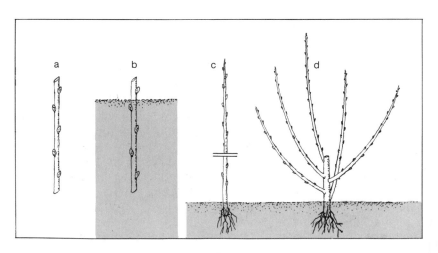

Currants

Where to Plant

Red and white currants are not necessarily undemanding, and they grow poorly in the shade or in poor soils. If grown under fruit trees, it is very important to provide adequate water and fertilizer, making sure that the needs of *both* tree and shrub have been attended to. Too, when growing berries in the shade you will have to reckon with higher acidity.

A deep, medium, fertile soil that is slightly acidic (pH 5.5–6) and free from weeds is best; only in such a situation will strong, heavily bearing branches develop.

The plants should have a protected situation, for late frosts can damage blossoms. If the plant is growing in an area so enclosed that air circulation is inhibited, plants such as gooseberries may have their fruit burnt by the sun.

Late frost damage on currant blossoms (*above*).

Sunburnt currants (*below*).

Planting

Because growth begins in very early spring, currants should be planted after the leaves fall in October or November. Fall planting also ensures good root and branch development in the spring. If you must plant in spring, do it as early as possible.

There has been a major change in the way planting is done. Bare-root planting, the former approach, could only be undertaken in spring and fall; today, with container-grown plants, planting can be done any time during the year except when the ground is frozen. Containers are large and similar to flowerpots, made of plastic or plastic bags of the sort in which shrubs, trees, conifers, and perennials are sold. Because the roots and rootball are not disturbed dur-

Currants

ing transplanting, container plants usually keep growing without a break. Place the container in a bucket of water, however, and keep it there until no more air bubbles rise to the surface. Finally, remove the container or tear open the plastic bag, and remove the plant. This sometimes requires a sharp knock to loosen it. If the rootball appears pot-bound—roots intertwined thickly all over the surface of the rootball—loosen them a bit and pull them out so that they will continue growth outward. Prepare the soil by loosening it and mixing in peat and compost. Set the plant in the hole with the rootball slightly below ground level. Fill in the hole and dig a trench around the plant. Don't forget to water during dry weather.

When planting in rows, keep the distance between rows wider than between plants. For moderate growers and white varieties, it should be about 6 × 4 feet, and 7–8 × 7 feet for stronger-growing types. Distances can be a little less if you do not plan to use a tractor. When using currants as a wind break, bushes can be planted more closely together. Plant espaliered or hedge plants from 10 inches to 3 feet apart.

Plant robust young bushes with three to six branches. More branches are unnecessary, because a maximum of six are allowed to grow. You'll remove the extra growths anyway so why pay for them?

In any case, thorough preparation of the planting area is necessary. Dig down two to three times the depth of the spade. However, only when planting a single tree or bush should you dig out a hole (with many trees, you'd use a channel). The hole should be wider than deep. Mix the soil from the hole with several shovels full of wet peat; this will hold moisture and stimulate the growth of roots. Well-rotted compost, if you have it, is excellent for use in planting. Cut back any damaged roots to healthy tissue. With a shovel—or by hand— refill the hole so that there are no air spaces. Finally, gently firm the earth around the plant by pressing it down lightly with your toes.

After planting, the bushes should be deeper in the earth than they were in the nursery. The results of planting too deeply are numerous. Plants grow only weakly. Regular watering is fine but not necessary in wet soils. What is important is covering the plant's "shoulders" (the ground around it) with a layer of straw, manure, or something similar to keep it from drying out. Shallow-rooted plants like currants benefit greatly from this type of mulching.

Currants

Tools for pruning: a bow-type pruning saw (*a*), heavy duty shears (*b, d*), a straight-bladed knife (*c*).

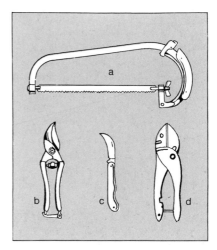

Pruning

Regular pruning may not be neglected, and its importance in maintaining fruitful plants not overlooked. Literature on growing berries contains page-long descriptions of pruning that sometimes irritate the reader more than enlighten. They talk about after-planting pruning, first-year pruning, second-year pruning, pruning the fruit-bearing limbs, and rejuvenation. It so happens that not all plants respond the same way to pruning, making hard and fast rules worthless.

There isn't room here to discuss pruning in all of its many forms, but it isn't necessary either. Bushes will continue to bear without your spending time and effort. Perhaps the berries will be slightly smaller or the bushes may bear less fruit, but the difference in yield may not be worth the effort. The same holds true for the usual Dutch method of growing berries commercially. The quality is excellent but requires wire fencing to which the shoots are trained and intensively pruned.

The following instructions for pruning are simplified. The goal of pruning in the first few years is to produce a bush with five to six branches that have as many short fruiting spurs as possible. The bush should be well filled out, but loose and airy.

Thinning

The basic scaffolding is built by pruning; pruning is important. Choose five or six of the best branches, and cut them back by a third, and then remove the rest. The reason for cutting back is that the end buds will be stimulated to develop fully and fruiting spurs will

Currants

be stimulated to grow from the many buds along the remaining branch. If the plant doesn't have five or six branches yet, the ones it does have should be cut back very hard to two or three buds to stimulate growth.

Training
Follow the directions for initial pruning for two to three years. The leaders are shortened every winter in order that more buds may develop into short fruiting spurs. There is no exact number of inches to be cut back. Here is where you have to use your intuition a bit and also observe how the plant reacted to last year's pruning. If it was cut back too hard, undesirable new woody growth will develop or there

may be equally undesirable bare spots. Cutting back approximately two-thirds of the new growth seems to be about right. If as a result of too vigorous cutting back the framework develops powerful woody growth, it will have to be shortened as well. Otherwise the crown will grow too thick. It should stay loose and airy in order to carry fruit on the inner branches. A healthy, fully grown shrub should have no more than eight or then branches.

Take care when shortening the branches, and always cut back to an outward-pointing bud to ensure a broadly pyramidal habit. The cut should be just far enough above the bud to avoid drying it out.

Other Reasons to Prune
Only after several years, when the bush becomes so dense that growth in the center becomes sparse, should further pruning be undertaken. This should be done by removing a few strong branches

Pruning a currant bush.

Pruning to rejuvenate an old currant that has grown too dense.

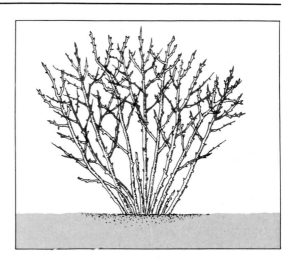

that grow toward the center, along with any suckers; suckers are useful only after about five years or so when the oldest branches producing the least fruit are replaced by a new framework. Also useful in forming a new framework are the branches that grow slightly above the ground from old wood. If you prune the original framework regularly, you can let it go a bit longer before replacing it.

The pruning operations described here do not require too much work and can be done quickly and easily. If the more difficult operations of thinning and rejuvenation are undertaken with an accompanying generous and specialized fertilization, one can expect a long plant life—twenty years or more—with ample harvests for the duration.

Standards
Pruning a standard currant is basically the same process. Five to eight leaders are cut back to four to six buds, with their laterals cut back to two to four buds. Thinning the crown and pruning for fruiting spurs are necessary as follow-up procedures.

Training on Wire
Commercial growers use a trellis of three wires at about 20, 42, and 65 inches in height, respectively. Plants are grown at 9-inch intervals in rows about 9 feet apart. The three branches of the framework of each plant are cut back every year,

Currants

to stimulate the growth of laterals; two weeks before the berries ripen, these side branches are cut back to eliminate shade. The remaining branches develop strong buds, ensuring the best-quality fruit for the following year. Pruning to stimulate fruiting spurs is extremely important when training plants on a trellis. The advantages of growing in this manner are quick, high yields that ripen at the same time because of even sunshine distribution, better-quality berries, and easier ground maintenance.

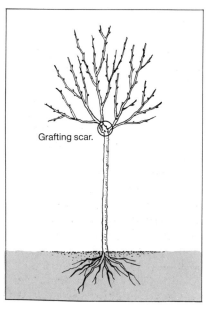

Grafting scar.

The Spindle System

A whole new method of growing red currants has recently been developed—a single-branch, pillar-style structure that is considered an improvement over training on trellises. Higher yields, better-quality fruit, easier picking, and more profitable cultivation are the goals of this method.

An absolutely vertical single branch is grown in the pillar system (often used with apples). It has the advantage of being able to grow laterals in all directions. These can be evenly covered with fruiting spurs. Eventually branches—with the exception of the central vertical—will grow more or less diagonally. What happens is that buds on the top of the branch develop strongly, whereas those below grow only weak side-shoots, or don't grow at all.

In the spindle-system plants obtained from the nursery with several branches are trimmed back to a single central branch, and one

Standard currant should be regularly pruned.

Currants

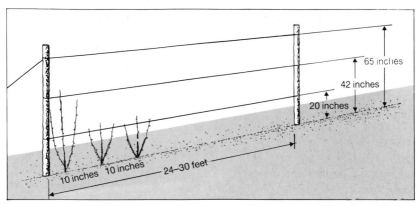

65 inches
42 inches
20 inches

10 inches 10 inches 24–30 feet

Training currants on a trellis. The three branches before pruning (a) and after pruning (b).

can use the cut-away branches for cuttings to grow more plants.

The remaining central branch develops quickly into a strong leader garnished with laterals. It can reach about 5 feet in height. In the second year, the laterals bear fruit, while the leader may reach 6 feet. The leader produces new laterals that will bear fruit in the following year. The laterals in the lower part of the plant develop branches that bloom the following year. Laterals must be cut back in winter, in order to prevent too many branches from developing, thus diminishing the quality of the fruit.

In order to keep growth around the axis as horizontal as possible,

cutting back of laterals should be done to outward- and downward-pointing buds. We know from experience with growing apples and pears in the pillar system that the horizontal position of the branches stimulates flower and fruit production. The physiological reasons for flower production and growth inhibition work with berry plants as well.

This short introduction to a new method of growing, which can also be used with gooseberries, will need to suffice. Gooseberries may be even better subjects for this method because their laterals grow easily into the desired horizontal position. In addition, anything that makes picking easier will be most

Soil

The traditional way of caring for the soil around plants was repeated hoeing to reduce transpiration and keep down weeds by stimulating capillary action. This is especially important, because berry bushes growing in weeds or grass do not thrive and consequently age quickly. The soil around the bush should never be worked deeply with a spade or even a rake, as the roots may be struck and damaged.

A better and more modern method of maintaining the soil is to cover it to a depth of 6 inches with a mulch of manure, peat, straw, or similar material. Such mulching has the following advantages: The soil under the mulch stays moist evenly, has excellent tilth, and becomes thoroughly penetrated with roots. Even after a hard rain there is no muddy soil, and you can always walk on it. Best of all the soil stays virtually free of weeds and the few that do manage to

noticeable when working with these prickly specimens. However, this new method was developed primarily for the commercial production of currants, and it requires pruning and a little practical experience. Whether or not it delivers all that it promises remains to be seen, but for advanced and interested gardeners it is worth trying.

penetrate the mulch layer can be pulled out easily.

Once in a while, loose layers of mulch harbor mice, but these generally are the harmless field mice kind. The much more damaging root-eating moles are no more numerous with mulch than without mulch. The only thing one has to watch for is the eventual decomposition of the mulch, which requires continual replacement.

Fertilizers

Fertilizing has two purposes: The first is to create favorable conditions for the vast population of microorganisms in the soil—that

Straw mulch under currant bushes.

Currants

is, maintaining the natural biological balance to ensure good tilth, the ideal condition for soil. It is up to the gardener to improve the physical structure of the soil so that microorganisms do their part in aiding decomposition—turning nutrients into forms that plants can utilize. Using organic fertilizers such as manure, compost, peat, and similar products enriches the soil and lets microorganisms begin their task.

A second reason for fertilizing is to quickly feed easily ingested nutrients in mineral form to plants. Dosage needs to be carefully considered, measured, and well balanced—because you can provide too much of a good thing. Balanced fertilizing means supplying all primary mineral needs (nitrogen, phosphorus, potassium, magnesium) and trace elements (borum, manganese, iron, zinc, etc.) in suitable combinations, and only if a deficiency is apparent should supplemental nutrients be administered.

With any deficiency, proceed with extreme caution. Take a soil sample to your local nursery and have both the mineral content and pH value determined, and have the leaves analyzed. The results will help you determine how to fertilize best. This is the approach with commercial fruit production, where large areas are considered, and it

is as important for the home gardener to keep informed of soil composition with occasional testing.

The following instructions may help determine the quantities of humus and manure necessary:

Organic Manures to Improve Soil

Manure.   Use roughly 4–6 hundred pounds per 100 square yards. Manure is also a useful top-dressing.

Compost.   Every two or three years add 2 cubic yards per 100 square yards of garden. Work into the soil before planting, or top-dress existing plants. You can't possibly add too much.

Peat.   Every three or four years add 2–3 bales per 100 square yards of garden.

Enriched Peat.   This is peat to which fertilizer has been added. There are a number of products available that enhance the value of peat, but they are also more expensive. Before planting add 1–2 bales per 100 square yards. Later (after initial fertilization), 2–3 bales per 100 yards.

Potassium deficiency shows up on currant leaves.

Chlorosis caused by iron deficiency; *right*, a healthy leaf.

Mineral Fertilizers

The following quantities per 120 square yards are recommended for berry crops:

In early spring, 8 pounds of slow-acting sulfate of ammonia or 8 pounds of quicker-working ammonium nitrate with 8 pounds of superphosphate. If you choose to apply in the winter, use long- and slow-acting bone meal. In addition, apply 8 pounds of calcium sulfate.

For heavy soils, gypsum may be added in winter or early spring to improve tilth; every third spring, add 35 to 45 pounds of ground limestone.

Complete Fertilizers

The easiest way to fertilize is to use what is called a complete fertilizer, containing the most necessary minerals and the most important trace elements in a balanced mix-

ture. Only use fertilizers that are free of chloride salts.

There are numerous fertilizers on the market in which the composition is nitrogen 12 percent, phosphorus 12 percent, and 20 percent potassium. Generally, there is an additional 2 percent of magnesium and trace elements as well. These proportions fulfill the needs of most garden plants. For two- to four-year-old red and white currants, use 10–15 pounds per 100 square yards. For bearing plants, use 18–25 pounds per 100 square yards. Otherwise, follow the manufacturer's instructions.

Quantities should not be given in a single but a number of applications. Begin carefully several weeks after planting, when the new growth is only an inch or so long. Use a handful (less than a cup) per bush, spreading it around the plant and working it in, and repeat one or

Currants

Currants

two more times at four-week intervals. Plants will respond with good growth. In subsequent years, you can gradually increase the quantities until the bushes are bearing fruit. At that time, make three yearly applications —twice in spring, and once following harvest to encourage the following year's bud development.

Generally it's better not to follow a rigid schedule when fertilizing: Let the plants "speak." Robust shoots and dark green foliage tell you the plant is in good health, and recommended applications can be decreased, while yellow, sparse foliage is a sign of hunger. In any case, don't fertilize too late in the year, and make sure that new growth has time to mature before the first frost.

Red currant Heros grown on a trellis.

Diseases and Pests

Die Back
It affects red and white currants, and black currants and gooseberries less often. Leaves show numerous large and small brown spots; leaves curl inward, turn yellow, and fall off; by the end of August, branches are bare. The cause is the fungus *Drepanopeziza ribis.* Spray with fungicide one to two times after flowering and again after harvesting. Similar symptoms are caused by the Fourlined Plant Bug, *Poecilocapsus lineatus.* If this is the case, plants will recover without help.

Nectria Canker or Coral Spot
Caused by the fungus *Nectria cinnabarina,* red pustules or cankers form on the canes of berry plants. The fungus enters the host plant through a wound caused by frost damage, insect bites, or pruning. It penetrates the wood and emits toxins that cause wilting and finally death. It is more prevalent in dry places.

Numerous light red cankers, carrying masses of ripening spores, appear on the dead plant parts. Often, for no apparent reason the fungus will suddenly stop growing so that it is possible to save the plant by cutting back to healthy wood. The diseased portions should be burned immedi-

Currants

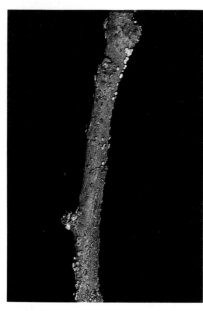

Red pustules on a currant twig caused by *Cryptomyzus ribis.*

ately and not used as mulch. There are no known fungicides that have proven effective.

Aphids
Crumpled, rolled-up leaves and crippled branches are the signs; look at the underside of leaves and on the tips of branches for numer-

ous light green, waxy aphids. The most common is the *Aphis schneideri.* They leave a trail of honeydew that supports fungi, and which renders the fruit worthless. To control them, destroy eggs in winter or very early spring with dormant oil, and for serious infestations use systemic rotenone.

Greenfly
Puffy red spots on red currant leaves, and on leaf undersides a few yellow-green aphids. The cause: *Cryptomyzus ribis.* Returns in late autumn to lay eggs; destroy with a winter or early spring spray.

Scale Insects
Scale insects are typical parasites of weakened plants: They are generally found on bushes in poor locations or on those which have been poorly nourished. Commonly seen is the European fruit scale or *Lecanium* scale, *Lecanium corni.* Nymphs winter under branches

Leaf damage; again, *Cryptomyzus ribis.*

Lecanium scale on a cane.

stems of leaves, canes, and fruits, red, pea-sized spots are found in the middle of which the insects appear as small gray scales. Later ash gray crusty areas composed of thousands of scales appear. Plants die. If you suspect an infestation, notify your extension service.

Red Spider or Spider Mites
Red spider mites (*Tetranchyus urticae*) have numerous hosts. Unlike other mites, they spin fine webs on the undersides of leaves and can move through the air suspended on fine threads. Of their seven annual broods, most live on nettles and other weeds. Their leaf feeding sites show up as specks, and later as a browning of the leaves, which eventually drop. To control you must take immediate action. Controls that *don't* destroy spider mites but help include spraying with cold water, a slurry of wheat flour, buttermilk and water, and natural predators among the beetles (*Stethorus picipes,* the spider mite destroyer). Alternatively malathion, demthoate, or derris help control red spider mite.

without their protective shield; in spring they migrate to a site, lose their legs, and remain fixed, becoming large, shiny brown, round scales. Abundant honeydew gives rise to fungi. Control: Spray in early spring on a sunny, warm day to destroy the wandering immature scale insects. Strengthen plants through fertilizing, watering, and possibly pruning.

The most dangerous scale insect is the San Jose, *Quadraspidiotus perniciosus,* common in warmer climates. Its preferred host plants include red currants. On the

Currants

Damage to currant leaves caused by the feeding of *Lygus pabulinus.*

Insect Damage

Very often shriveled, crumpled, and torn leaves are found on currants, the result of the tarnished-plant insects, *Lygus lineolaris* and *Lygus pabulinus,* feeding on the plant and leaving poisonous spittle behind. To control: Spraying in spring may help, but isn't necessary, and sabadilla dust for more serious cases.

Black Currants

From a nutritional standpoint, the black currant is without doubt our most valuable berry. Not only is it excellent in the preparation of costly confitures, but the milder-flavored varieties are excellent for eating fresh. Its best use is juice that is so nourishing it is almost curative.

Black currants have an unusually high level of vitamin C that can reach 200–300 mg per ¾ cup (100 g). But it is not only this high vitamin content, in combination with important minerals like calcium as well as citric and tartaric acids, that make the black currant so valuable they also contain a high amount of the so-called *P* factors—*P* standing for permeability, which improves circulation in the capillaries.

The nutritive value of black currants affects the overall mechanism of important substances in the human body, and it is little wonder that during the war black currant juice was used medicinally, reserved for children, old people, the sick and the hospitalized.

Currants

Botany

The botanical name of black currants is *Ribes nigrum;* it is native to Central and Eastern Europe as well as a number of Asian countries. Its branches grow lustily and upright, and are typically golden brown with downy new shoots. The leaves are round, with three lobes, and the yellowish glands on their undersides emit a distinctly unpleasant smell. Greenish flowers with pale red centers hang in clusters. Their botanical interest is that they run the whole gamut, from totally self-fertile types (Roodknop) to the virtually self-sterile (Rosenthals Langtraubige). Planting several varieties together in proximity to bees, which do most of the pollinating, enhances the harvest. Increased berry size results when large numbers of seeds are developed.

Varieties

Silvergieters Schwarze.   This is an early bearing Dutch species (1936). It isn't fussy about its situation, grows quickly, upright. Very long clusters of big, deep blue-black berries a little less sour than other varieties, therefore sweet and mild tasting. Good for eating fresh and/or juice, jams, and wine.

One of the oldest species Silvergieters Schwarze.

Rosenthals Langtraubige Schwarze.   (Synonym, Boskoop Giant.) Early ripening species (1913) that demands good situation and soil. Strong spreading growth. Large harvests despite partial sterility. Long clusters of big, thin-skinned, sour, aromatic berries. It is highly prized commercially because of its high mineral and vitamin C content.

Goliath.   Early. Especially adapted for heavy, moist soils. Not sensitive to cold. Strong, spreading growth. Fruits carried close together, and

Currants

Black currant Goliath.

various-sized berries are difficult to pick. For this reason, and also because of the staggered ripening, not grown commercially, but its pleasantly mild, bittersweet taste recommends it for fresh eating.

Roodknop. An old Dutch species (1921). Bears early. Strong growth on broadly rounded bushes; easily grows too dense. Blooms late. Good, dependable harvest. Short to medium clusters of uniformly large, soft-skinned berries. High mineral and vitamin C content. Good for canning despite its undesirable brown juice.

Wellington XXX. English species (1951). Early. Strong, broadly round bush with good branching. Yearly thinning is necessary. Long clusters with short stems of big, thick-skinned, aromatic berries are good for eating fresh. High mineral and vitamin C content.

Baldwin Hilltop. Old English, late-ripening species. Thrives in a moist spot. Moderate growth of widely upright bush; medium clusters of closely spaced fruit. Berries large, thick-skinned, and good tasting. Very high mineral and vitamin C content (230–300 mg per 100 g).

Daniels September. Old English species (1923). Late to very late. Not suitable for heavy soils. Strong growth, broadly round, drooping; structural pruning necessary. Abundant, reliable harvests. Long fruiting clusters are loose with medium-sized, firm, aromatic berries; mineral and vitamin C content high.

Currants

Propagation

Propagation of black currants, like that of red and white, is often left to nurseries. The following circumstance should, however, be noted: Plants of the same species growing in our gardens are unfortunately not always uniform. Often, different mutant characteristics arise. Some may have better (or poorer) harvests than normal, and it is only natural to wish to propagate the more bountiful variations. Because propagating black currants is quite simple, propagating a superior variety is doubly worthwhile. Directions are the same as for the red and white (*see* page 16).

Both commercial grower and home gardener should make a habit of observing their plants' characteristics—vegetative and generative performance, as well as any positive variations through mutation that may make them good candidates for home propagation—always a better choice than planting an unknown quantity.

Where to Plant

The most important consideration in planting is finding a sunny, protected spot for the highly frost-sensitive flowers. Gardens in frost hollows and dips in the ground where frost gathers with no place for the cold to escape are not suitable. The soil should be somewhat heavier, richer, and cooler than those for red currants. Sandy soils can be used if they are enriched with humus and nutrients and kept moist.

Frost crack on a black currant.

Currants

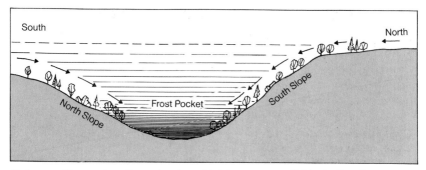

Hollows and low spots, with no place for cold air to go, are unsuitable planting sites.

Planting

The procedure for planting black currants is the same as for red. Because the black are stronger growers, leave more space between bushes: 8 to 10 feet between rows, and 5 to 8 feet within the row. A single plant should have a minimum of 5 to 8 feet of space all around. Bushes are planted (unlike red currants) 3 to 4 inches deeper than in the nursery in order to encourage branching for crown development and to delay renovation. (These distances are for black currants growing as bushes; standards and plants trained to trellises require less space.)

Standards

Frequently currants are trained as standards. While commercial growers use this approach because the berries can be picked by machine, the home gardener chooses it because it makes caring for the area around the plant easier and likewise facilitates picking.

To grow a standard, you need the yellow currant, *Ribes aureum,* of a variety like Brechts Erfolg for root stock, and a scion of any variety that lends itself to this form. Graft together at the desired height—anywhere from 1 to 4 feet.

For stability and wind and storm protection, stake the plant—preferably with a support that reaches into the crown. Any suckers or laterals of the *Ribes aureum* rootstock should be removed as soon as they appear. Standards should be spaced about 3 feet apart. Just as it is with red currants, training on trellises is another possibility. Proceed with planting as for red currants (*see* pages 17–18).

Currants

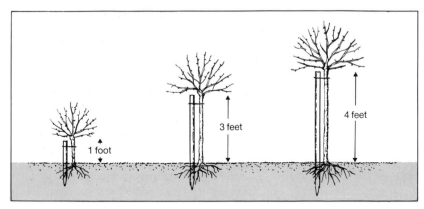

Black currant standards with different-sized stems. They require staking.

Pruning

Pruning of black currants is similar to that for red and white. The goal is a loose, airy crown composed of five or six evenly spaced branches. When pruning, remove weak growth and cut back the branches that will be kept. If the bush doesn't have enough branches, cut back hard the ones it does—to two or three buds—to stimulate new growth for the development of a crown.

Thereafter, remove all branches that are too close together, all side branches below the place where you cut back, any branches that grow inward, and the extras growing out of the ground. Eight structural branches are the maximum for a mature bush. Remember that black currants bloom on one-year-old wood. That means that we have to cut back for at least two years to produce a bush with strong one-year growth and, above all, lateral branches.

Currants

When the old framework of branches ceases vigorous growth (after about four or five years), it is replaced with new growth. Strong new shoots emerging from the ground are cut back to encourage growth and used in place of the old framework, which is removed. Sometimes another branch—not from the ground but growing on the old scaffolding—can serve as a replacement. The continual renewing of the old wood and moderate thinning are the two aims of pruning black currants. When pruning, take the time to observe the plant's response to the previous year's pruning.

Naturally, if you are willing to extend the effort, black currants may be grown on trellises or as standards. Use the same procedure as for red currants (*see* pages 21–24).

Soil and Fertilizers

Care for the ground under your plants in the same manner as described for red currants (*see* pages 24–25). They have the same needs for moisture and humus, and will also benefit from mulching—the covering of the ground with slowly decomposing organic material. Of course enriching the soil with organic material is always beneficial.

It is the same with chemical fertilizers. The most important nutrient to provide is nitrogen—especially because the continual development of new shoots in all parts of the foliage makes heavy demands.

Thinning black currant bushes.
Before thinning.

The same bush after thinning.

Currants

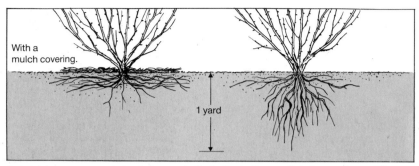

A mulch layer encourages roots to grow in the fertile, better-ventilated upper layer of the soil.

Black currants need about 25 percent more nitrogen than red. Mature shrubs benefit from the application of ½ ounce (15 g) of pure nitrogen per square yard. Again, it is best to divide the total amount to be given into two doses: two-thirds in early spring, and one-third in May after fruit has set. Should fruit be absent due to frost damage, this last dose can be dispensed with. Experiments carried out on black currants show that nitrogen in nitrate form works better than in an ammonia form. Otherwise, the nutritive needs are the same as those for the red.

Diseases and Pests

Fruit Drop
This sickness has no parasitic cause. A few weeks after bloom, the young fruits of black currants drop off—especially from plants with long clusters. Occasionally this also occurs in red currants and even gooseberries. A number of factors, aside from genetic inclination, may be responsible. Most often damage to flowers caused by late frosts is to blame, or inade-

Fruit drop, often caused by late frost damage.

quate fertilization due to cold, wet weather, or too few bees or even poor pollination. The more seeds that develop in a fruit, the more growth is stimulated, which is important for fruit development. Otherwise the corky tissue growth forms too soon, causing the fruit to detach before it is ripe.

White Pine Blister Rust

First, thousands of tiny orange (or yellow) pustules appear on the bottom of older leaves. Leaves drop early. The cause is the fungus *Cronartium ribicola,* which uses the Eastern white pine, *Pinus strobus,* as a host for another stage. Blister rust is so serious for pines that some states do not allow currants or gooseberries in. To control: Two to three preventive sprays with fungicide (Zineb), once right after blooming, and once or twice after harvesting.

Currant Bud Mite

Sometimes this mite is also found on red currants and gooseberries. The buds swell into balls. Adult form of the currant borer that hollows out canes; they don't develop and dry up. Inside tiny white mites of the species,

Black currant blister rust. Spores appear on the undersides of leaves.

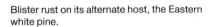

Blister rust on its alternate host, the Eastern white pine.

A serious case of currant bud mites showing numerous rounded buds.

Currant Borer

A moth with wasp-like appearance (*Syanthedon tipuliformis*) lays eggs in May and June. Young borers penetrate the canes (often through pruning injuries and buds) and hollow them out. Above the place where they entered, growth dies. Loss of canes continues. White worms between ¼- and ½-inch long winter in the plant and pupate in spring. It is difficult to control chemically because moths don't stay in place. Control: Possible insecticide spraying at the end of May, beginning of June. Burn all affected canes.

The currant borer hallowing out young shoot.

Cecidophyopsis ribis may number to 10,000 within a single bud. Control: If there are only a few rounded buds, remove them; for heavier infestations remove all stricken parts of the plant. Several thorough sprayings at the end of March and early April will eradicate young wandering mites. Use a sulphur-based preparation. This dangerous pest can pave the way for other disease.

Currants

Reversion of black currant bush.

Reversion
Black currants (and occasionally red) sometimes grow distorted leaves, involuted individual lobes, and sharply toothed leaves. Flowers are sterile. The cause is a microplasm. These tiny, primitive organisms, which are ordered between viruses and bacteria, were first discovered in 1967. This condition often follows an infestation of currant bud mites. To control: Uproot and destroy the affected plants.

Gooseberries

Why must gooseberries stand in the shadow of other berries? They have been around longer, and many people prefer them to currants and other types of berries. The reasons are twofold: Harvesting is difficult because of their spines, and commercial possibilities are limited because they are not versatile. It is possible—with much sugar—to make juices and jams from gooseberries, but the products are simply not as good as those made from currants, raspberries, blackberries, and strawberries. Nevertheless, gooseberries in all of their different colors and tastes deserve great popularity and are special luxuries belonging in every garden.

Botany

Gooseberries are native to Europe, North Africa, and Asia from the Caucasus to China. They grow wild in the woods, becoming about 5 feet tall and beginning growth very early in the year. Our gooseberry varieties belong to the species *Ribes uva-crispa,* a variety of *sativum* (formerly called *Ribes grossularia*), which belongs to the family Saxifragaceae. Very often gooseberries are grown as grafted standards on the yellow currant rootstock *Ribes aureum,* Brechts Erfolg.

The generally three-lobed leaves are only a few inches across. Flowers are inconspicuous, single, yellow-green, or slightly pink in appearance, and grow in sparse clusters. They are protandrous, and the female organ, the pistil, is not ready to receive the pollen when it ripens. Therefore, it is absolutely necessary to plant several varieties with different blooming times, and to encourage the presence of bees.

Propagation

Like the currants, gooseberries are propagated vegetatively, not by seeding. Instead of taking cuttings, new plants are produced by ground layering or by encouraging branches to root by mounding earth over the plant. Propagation of cuttings or green wood should be left to nurseries that have all of the appropriate equipment on hand. New plants produced by ground or mound layering are quite easy to produce without any prior experience.

Ground layering can be done in spring when a strong, one-year-old branch is laid flat on soil that is enriched with peat or compost. It is fastened down and covered with earth. The buried branch will send out shoots from its nodes around

Gooseberries

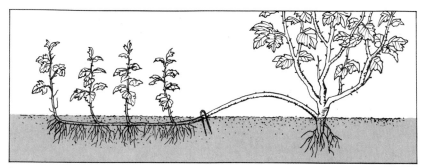

Propagation of a gooseberry plant by ground layering.

which compost or good, rich earth is heaped. The new plants are allowed to develop roots until fall. Then, the buried branch is divided so that each rooted shoot with its piece of branch may be grown as an independent plant.

Using the mound-layering method, in spring the mother plant is cut back to the ground. As the branches begin to grow, earth is heaped around the mother plant to encourage root development along the buried stems. In the fall, the earth is carefully removed with a garden fork and the rooted young shoots are cut off the mother plant. In general, they need a year to mature before they are set in their places in the garden.

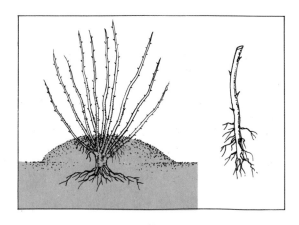

Mound layering a gooseberry plant. A rooted shoot (*right*).

Gooseberries

The first virus-free young plants were produced by meristem culture in the year 1968; today, many tissue-cultured varieties are available.

Varieties

Shakespeare (1564–1616), England's greatest author, wrote about gooseberries, and since that time England has been the country where gooseberries are most at home. In fact, most gooseberries grown around the world have English origins. As early as 1802, a German pomologist, J. E. Christ, described 290 English varieties as opposed to only eight from his own country; in 1913, L. Maurer, writing in his well-known *Gooseberry Book of the Best and Most Widely-Grown Gooseberries,* listed 133 red, green, and yellow varieties, of which most were English. Luckily, the assortment we find in nurseries today has shrunken to a sensible size. From the red, green, and yellow we have chosen the following varieties:

Red

Whinham's Industry. (Rote Triumph.) An old English variety. Early. Strong, broadly round heavily spined bush. High, even bearing. Medium to large, elliptical, dark red, thick, hairy-skinned berries. Bittersweet taste. Good for eating and canning. The most important red variety, robust and adaptable to site.

May Duke. (Maiherzog.) English variety. Early, medium growth, very dense with hanging branches. Thinning necessary. Medium harvest. Fruit is large, round, light to purple-red, thin-skinned, many thin branches. Medium harvest, somewhat uneven. Berries are medium-sized, red, thin-skinned, and susceptible to sunburn; they ripen unevenly. Highly prized for eating fresh and making jams. Very susceptible to gooseberry mildew and die back.

Crownprince. (Rote Orleans.) Mid-season to late. Moderately strong growth. Bush-high, round, few drooping branches. Harvest moderate. Berries are big, pear-shaped, dark red with scarcely any fuzz. A valuable, good-tasting variety for eating fresh. Generally healthy.

Gooseberry May Duke grown as a pillar.

Rote Preis. An old English variety that is very late. Medium growth, covered with drooping clusters; needs cutting back to stimulate growth of scaffolding. Medium harvest. Berries are very large, long, dark red when fully ripe, bittersweet, aromatic, good for preserving, and despite a somewhat thick skin, good for eating fresh. Generally healthy.

Yellow/Green

Hoenings Fruehste. German hybrid, grown since 1900. Very popular early yellow. Strong growth, stiffly upright; well-armed with long spines. Thinning is useful. Good harvest. Fruit is medium to large, round to oval, golden yellow, thin-skinned, with thick fuzz. Sweet, delicious. Not very susceptible to mildew. Not fussy about site, responds to good care.

Triumphant. (Gelbe Triumph.) An old variety (1889) of unknown origin. Mid-season. Growth strong to moderate, flat, drooping. Fruit is medium, long, oval, greenish yellow, often with red spots, sweet but not very aromatic. Better for picking green and preserving than for eating fresh. Not fussy about site.

Lauffener Gelbe. German variety (1938). Strong, spreading growth. Mid-season. Uneven harvests. Berries are medium-sized, uniform, golden yellow, delicious, with fine down. Susceptible to late frosts, mildew, needs a sheltered place. One of the best for eating fresh. Green-white varieties.

The highly prized, delicious-tasting yellow gooseberry Hoenings Frueheste.

Weisse Neckartal.   German variety (1942). Early to very early. Strong, upright growth with good branching. Good harvest. Berries are medium-sized, round, white to yellow-green; skin is thin. Sweet, aromatic, unusually good tasting, favorite eating variety. Unfortunately, somewhat susceptible to mildew.

Weisse Triumph.   (Gruene Hansa.) Old English variety (1802). Mid-season. Strong, upright growth, well-branched. Strong spines. Very large, regular harvest. Fruit large and round. Skin thin, white to yellow-green, with brown dots, light down. Fine, bittersweet, aromatic taste. Especially good for eating raw, but also useful for preserving. Very susceptible to mildew. Not fussy about climate or site.

Gruene Kugel.   Germany hybrid (1940). Mid-season. Strong growth, well-branched, big, tall bush. Very large harvest. Berries are large, round, light green, with thick almost smooth skin. Mild taste, not aromatic. Mildew.

Lady Delamere. English variety.
Late. Very strong growing, broad,
round, widely spreading. Thinning
is necessary. Bears well and regularly. Berries medium to large, long
ovals, light green with thin skin.
Bittersweet taste, not aromatic.
Often burns. Good for picking
green.

Where to Plant

The ideal site for gooseberries is
similar to that for currants. They,
too, need a bright sunny spot, but
not too much of a good thing.
Some varieties don't bear well and
are damaged by long, intensive
exposure to sun. The blossoms are
very sensitive to frost and like the
currants they may not bear some
years due to late frosts.
Gooseberries require a medium,
loamy, rich soil to be at their best;
dry, infertile, sandy soil doesn't suit
them and under these conditions
die back is common. Such soils
should be enriched with humus

Gooseberry Lady Delamere.

Sun damage on gooseberries.

Gooseberries

and nutrients, and watered during dry weather. The best site is a good, cool, humus soil in a bright spot out of the wind.

Bushes grow better with some type of mulch—even if gooseberry roots don't grow as close to the surface as those of currants.

Planting

Gooseberries require the same soil preparation, planting time, and planting procedure as currants. Spacing of gooseberry plants, like currants, should be wider between rows than between plants. Generally, the distances between gooseberries are slightly less than between currants—but it depends upon soil quality, growth rate of the variety, and pruning regimen. 5 to 7 feet between rows and 3 to 5 feet between plants is recommended. High standards, middle-sized, and low ones, often planted in gardens, should have at least 3 to 4 feet around them. A combination of bushes and standards is possible. Intervals of 8 feet between standards will allow for one gooseberry bush between.

Plant bushes the same or only slightly deeper than they were growing in the nursery; otherwise, they will exhaust themselves producing new shoots. Watering is recommended, except in moist soils. It is important to cover the ground around the plant with peat, manure, or a similar medium.

Pruning

Certain pruning has to be done to keep the gooseberry from quickly becoming a thick wilderness of branches. Neglected gooseberry stands often look this way.

Initial Pruning

Gooseberries require basically the same pruning and training as currants. Choose the strongest, healthiest, and best-positioned branches to build the scaffolding; cut back by one-third and remove all other branches. A full, mature bush should have six to eight evenly spaced branches.

Training and Pruning Established Plants

Gooseberries bear on wood grown the previous year. It is necessary to keep a supply of one-year-old wood growing. Regular, moderate pruning in order to stimulate the growth of side-shoots is undertaken every year. Laterals are also trimmed, although not as vigorously as with currants. The yield will be smaller than with currants, but the fruit will be bigger and more

Gooseberries

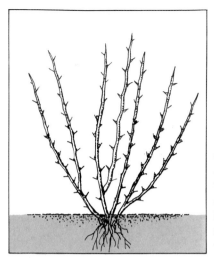

Cutting back a gooseberry bush.

perfect. Because mildew is the number-one problem of many gooseberry species, usually affecting the branch tips, regular pruning also functions as a preventive measure.

Another result of regular pruning is the formation of many buds, causing the crown to become too dense. Improving circulation and allowing light to reach all parts of the plant is an important function of pruning. Completely remove all branches that are too closely spaced, those that grow toward the center, extra stems emerging from the ground, and any branches that droop toward the earth. When drooping branches bear fruit, it is frequently mud-spattered. They also usurp the energy of horizontal or even upright branches. Another way of handling them is to retrain them by cutting back to an upward pointing bud—not the usual procedure for pruning.

Like currants, gooseberries need a rejuvenation every four to six years—that is, a new set of scaffolding branches. We can use vigorous new branches arising from the ground or branches from the lower part of the old scaffolding. Of course, gooseberries may also be trained to a trellis, as is commonly seen in Holland. To grow gooseberries in the pillar system, follow the instructions for currants on pages 22–24.

Gooseberries

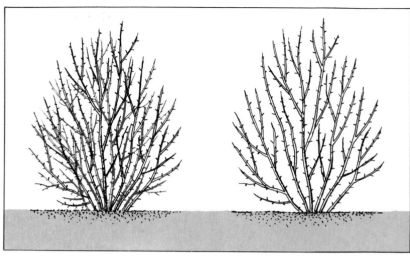

Thinning and rejuvenating an older gooseberry.

Soil and Fertilizers

Follow the same procedures for caring for the soil around the bushes and for mulching that are recommended for currants.

Gooseberries' nutritive needs are very nearly the same as those of currants. Apply 16–18 pounds of nitrogen per 100 square yards, or somewhat less if a nutrigen-rich mulch such as manure is used. Many complete fertilizers designated 12 percent nitrogen will supply 16–18 pounds of nitrogen per 100 square yards. Gooseberries need large amounts of potassium in a form free of chloride salts because of the plants' great sen-sitivity to chloride. Two-thirds of the amount of fertilizer to be used are given in spring, the rest in May—but only if the plant has set fruit. Otherwise this last dose can be bypassed.

Gooseberries

Harvest

Gooseberries are the only berry that can be harvested green. Early varieties may be picked from the middle of May on, when the berries are approximately one-third of their ultimate size. It goes without saying that these are for cooking and preserving only. In both cases large amounts of sugar must be added. Some varieties turn brown during preparation, which for commercial purposes is undesirable, but it shouldn't bother the home canner because this neither affects the shelf life nor taste of the finished product. Nearly every variety allows for harvesting over a period of several weeks, as opposed to a single harvest, and for the home gardener the advantage is that the ripest and best berries are picked each time.

Diseases and Pests

Gooseberry Mildew

The cause is the native fungus *Sphaerotheca mors uvae,* which occasionally attacks currants. A white (later brown) layer of fungus coats the branches' leaves and fruit, rendering the berries inedible. To control: Keep the bushes well thinned out to allow for good sun penetration and air circulation. Cut away and burn any diseased wood. If necessary give a late winter spraying; otherwise, apply a mildew fungicide three to five times after the flowers open. Make certain the soil is well limed and supplied with adequate amounts of phosphorus and potassium. Excessive nitrogen promotes the fungus. Thanks to successful crossing, good, mildew-resistant plants will be more and more available.

Gooseberries

Gooseberry mildew caused the tip of this shoot to die.

Fruit may also fall prey to mildew.

Die Back

The same problem that occurs with currants, so-called die back, can affect gooseberries. Caused by the fungus *Drepanopeziza ribis,* it usually develops immediately following harvest—often after a period of rainy weather. Different varieties are affected in various degrees but, in any case, loss of tissue harms the plant. Because many gooseberries are sensitive to copper sprays, use an organic fungicide such as Zineb.

Aphids

Most commonly found on gooseberries are the little currant aphids, *Aphis grossulariae*, which appear in dense colonies on all parts of the plants. Eggs laid on the branches to remain over the winter can be sprayed in winter. Otherwise, use a commercially available aphid spray or rotenone. Never use a systemic if you are considering picking the fruit green.

Gooseberries

Red Gooseberry Mite

This mite, *Bryobia ribis,* can do serious damage to gooseberries in the spring. The leaves turn a whitish gray and dry up from many hundreds of tiny bites. Bushes appear to have been burnt. Mites migrate in June. To control: Spray as soon as possible.

Gooseberry Sawfly

If plants are suddenly eaten bare during the summer, the cause is a ¾- to ⅛-inch-long caterpillar, the larvae of the gooseberry sawfly, *Pteronidea ribesii.* Generates severally. To control: Spray as soon as you see the pest.

Bird Damage

Often bullfinches, finches, and other birds will eat the buds. If necessary, the bushes may have to be either covered with netting, or sprayed on a dry day toward the end of December. Use a commercial antibird preparation. Sometimes growing alternate plants that provide feed for birds are helpful in keeping them away from the berry bushes.

Denuded gooseberry bush after an attack by *Pteronidea ribesii* caterpillars.

Jostaberries

Experimental crosses of different berries were begun as long ago as 1922. Professor Erwin Baur, of the one-time Institute for Genetic Study in Berlin, crossed a mildew-resistant wild North American species of Ribes (*R. succirubrum*) with a mildew-susceptible gooseberry species with large berries. The result was a strongly growing, richly bearing plant with clusters of fruit that he named Jochelbeere, combining the German common names for currant and gooseberry.

Interest in hybridizing was excited by Baur's work and carried further at the Kaiser Wilhelm Institute for Hybrid Research, founded in 1928 in Muencheberg, Germany. The goal was a new, better, and more richly bearing hybrid resistant to gooseberry mildew (*Sphaerotheca mors uvae*), die back (*Drepanopeziza ribis*), and blister rust (*Cronartium ribicola*). After forty years of crossing, a new kind of berry was born, jostaberries (*Ribes nidigrolaria*), from a cross of the black currant and the

The jostaberry (*left*), with its "parents," the black currant and the gooseberry. A twig with fruits (*right*).

Jostaberries

gooseberry and *Ribes divaricatum.*

The name Josta, from the German for both currants and gooseberries, is the trade name assigned to this variety by the hybridizer Dr. Rudolf Bauer, in 1975. The bushes grow stronger than either black currants or gooseberries, reaching over 5 feet by their second year. Like black currants, they have no thorns. Their hardiness approaches that of gooseberries.

The plants begin growth in very early spring. The shiny, dark green leaves have a size and shape midway between gooseberries and currants. In fall they remain on the bushes for some time. Noteworthy is their complete resistance to gooseberry mildew and the fungal cause of die back. Jostaberries are also fully resistant to the currant bud mite that causes round, swollen buds, and carries reversion and other viruses or microplasms of the black currant.

Flowers are larger than those of the black currant or the gooseberry, hanging in clusters of three to five on one-year-old wood. Older wood sprouts fruiting spurs, like the gooseberry, so that the bush bears for many years in all parts. The jostaberries won't develop bare spots as easily as the black currant. They bloom at the same time as their parents, but bear a little sooner.

Berries are bigger than black currants but smaller than the large gooseberries. They are elliptical to oval in shape and resemble the green stage of gooseberries. Like gooseberries, they have a little stalk which the black currant does not.

Ripe berries are dark with a smooth skin without the oil glands that give black currants their characteristic odor. Both berries and clusters hang firmly on the bush even when fully ripe.

The taste of jostaberries is totally new—a combination of the fine taste and aroma of gooseberry, with a bit of the black currant mixed in. They are extraordinarily high in vitamin C (90–100 mg), some of the richest of all fruits. The berries are excellent for canning, freezing, and they make delicious, aromatic jams and juices.

The jostaberry harvest far exceeds that of the black currant. Averages over years indicate a yield of approximately 11½ pounds per bush. In years with late-spring frosts, jostaberries, like gooseberries and black currants, sometimes won't produce a harvest.

Jostaberries

Branches of jostaberries, with their rich harvest.

Planting and Care

Each bush should occupy 4 square yards. Cutting back to stimulate growth is not necessary because the bush grows vigorously and branches well. In its third year, the bush will have reached a physiological balance between vegetative and generative development, and will remain so. Thinning is only necessary if vigor is markedly reduced; then, fertilize like black currants with a complete fertilizer. Pruning can be restricted to a moderate thinning of branches that have grown too thick. Most often trimming drooping branches is sufficient. Rotating in new wood for old structural branches isn't as important as with black currants.

Because of its built-in resistance to mildew, die back, and mites, sprays are rarely needed. Aphids could present a problem and can be treated with a commercial preparation.

Jostaberries are ideal for the home or market gardener: They are vigorous, thornless, and relatively untroubled by insects and diseases, and provide rich harvests of luxurious fruit. For these reasons they can be grown instead of or beside the black currant, but they should be grown nonetheless. Raspberries are garden treasures. Is there anyone who doesn't prize them? Yet only infrequently do we find small quantities offered for sale. The reason is the high cost of picking and transporting the fruit,

Raspberries

factors that also contribute to their exorbitant price.

Raspberries are one of the most reliable fruits. Late frosts are not a problem. They quickly produce bountiful harvests. Their unusual popularity arises from their versatility: Not only are they marvelous for eating raw, they are equally good in compotes, jams, jellies, juice, sorbets, on fancy cookies, even in the preparation of liqueurs, and several species are excellent for freezing (Zefa 2, Multiraspa, Himbostar). Now that perfectly healthy plant material with natural immunity can be obtained through tissue culture, raspberries belong in every garden. It's impossible to have too many.

Botany

The botanical name for raspberry is *Rubus idaeus;* it's a member of the rose family. The species name, *idaeus,* goes back to Pliny's notion that raspberries were medicinal plants that came from Mount Ida in Greece. The species-rich genus *Rubus* encompasses biennial shrubs—even low-growing, herbaceous types. Among these are *Rubus arcticus,* the Arctic blackberry, and *Rubus chamaemorus,* the cloudberry, both of which are native to the northernmost regions of the earth.

Unlike true shrubs that keep their scaffolding branches for many years, raspberries are perennial plants with biennial stems. The young canes grow from the end of June until about the middle of September, bloom, bear fruit the next year, then die. They are self-fertile, although it is still advisable to plant several varieties because cross-pollination always increases both harvest and berry size. While most raspberries bear only once, there are a few that yield two harvests, bearing in fall on new canes. Most raspberries are red; only a few have pink or yellow berries, but these are infrequently found.

Raspberries

Last year's canes (*right*) bring in this year's harvest. Young canes for next year's harvest (*left*).

Propagation

Producing young raspberry plants is extremely easy, but really not recommended because so many mother plants may harbor viruses. Simple propagation requires only transplanting the strongest of the rooted shoots into either a nursery bed for a year or directly into the garden. But only take shoots from a mother plant that is vigorous, bountiful, true to species, and absolutely healthy and virus-free. It is always better to leave propagation to a nursery whose certified plant material and controlled conditions will ensure vigorous and absolutely healthy young plants.

Another way of obtaining virus-free raspberry plants is with the relatively new meristem culture developed in 1970, or through thermoculture—a process with great promise for the future. Here rooted shoots are grown at temperatures between 98 and 100 degrees F, after which the tips of the shoots are removed, protected against insects (which carry viruses), rooted, and propagated. Since 1981, research on strawberries using a combination of thermo and meristem cultures is being carried out by R. Hummel of Stuttgart/Weilimdorf, Germany. This method may prove to be the best of all for producing virus-free plants.

Naturally, both meristem and thermoculture have to be conducted in special laboratories by trained technicians.

To date, experiences with tissue-cultured raspberry plants are overwhelmingly positive and promising. Experiments carried out at German horticultural institutes in

Raspberries

which tissue-cultured plants were compared with vegetatively produced raspberries resulted in double the harvest in the meristem plants. One variety, Schoenemann, actually produced a harvest fourfold greater. In addition, the meristem plants were markedly larger and more robust. This modern propagation method not only eliminates the all-too-common raspberry viruses, it also allows for mass production of valuable mother plants or new varieties in short supply. And it can be employed at any time of year.

Varieties

Glen Clova.   Scottish variety (1969). Very early to early. Strong to moderate growth, stiffly upright. Produces many runners. Fruit is medium to large, longish, round. Fairly firm, light to orange-red in color, and aromatic.

Malling Promise.   English variety (1944). Very early. Growth moderate, drooping, produces many rooted shoots in a wide range. Relatively long medium-sized harvest. Berries are big, round to bullet-shaped, light red, soft, sweetly aromatic. Plants are somewhat sensitive to rain, moisture, and wind, but winter hardy. Not recommended for freezing.

Malling Exploit.   Sister variety of Malling Promise (1944). Early. Growth moderate to robust, drooping, with many runners. Berries big, conical, light to medium red, strongly aromatic. There is a tendency for fruit to crumble. Extremely hardy.

Malling Jewel.   English variety (1950). Early. Especially suited to a maritime climate and for wet years. Widely grown in England. Upright growth, stiff with few runners. Berries are medium, conical, mid to dark red, somewhat dull, moderately firm, good tasting.

Raspberry Malling Promise.

Raspberries

<u>Multiraspa.</u> Hybridizer, Professor R. von Sengbusch (1971). Mid-season. Growth moderate to robust; moderate production of rooted shoots. Notable harvests. Berries big, round to conical, light red, firm. Sensitive to drought.

<u>Zefa 1.</u> Hybrid of the Confederation's Research Institute, Waedenswil, Switzerland (1960). Mid-season. Robust growth, many young canes. Big, conical berries, medium red, firm, good tasting. Very hardy, suitable for growing at higher altitudes.

<u>Zefa 2.</u> Another Swiss hybrid, Waedenswil (1960). More robust than Zefa 1. Mid-season. Strong to medium growth, with vigorous young canes. Berries medium to large, roundly conical, dark red, firm; taste is tart, aromatic. For less-than-ideal soils with poor substrata.

From top to bottom: The varieties Zefa 1, Zefa 2, Zefa-Herbsternte, Schoenemann.

Raspberries

Himbostar. Swiss variety (1975). Late. Growth is moderate, few young shoots. Berries large, conical, light red, firm, good tasting and tart. Needs wind protection and adequate watering.

Schoenemann. German variety (W. Schoenemann, Fellbach/ Wttbg., 1950). Good late variety. Vigorous growth, with many rooted shoots. Better-than-average harvest. Berries are large to medium, long conical, dark red, firm, tart, and good tasting. Very popular. Suitable for preserving. Has been greatly improved through selection and tissue culture.

Zefa Herbsternte. Twice-bearing raspberries are infrequently grown because harvests are too small; nevertheless, very nice to have raspberries in September. Ripening continues over a long period, until the first frost. This hybrid from Waedenswil, Switzerland, is the best twice-bearing raspberry today. It prefers a warm, sunny setting. Moderate growth, canes only about 3 feet long. Berries are medium, conical, bright red, good tasting. Main harvest is in September.

Other Varieties
Notable new hybrids from recent years are: Malling Orion and Malling Admiral from England, Sirius and Spica from Holland, Veten from Norway, Glen Isla from Scotland, Bulgarski Rubin from Bulgaria, and Gigant, Rutrago, Rucami, Rumilo and the double-bearing Korbfueller from Germany. Of these hybrids some will become famous and remain, others will disappear. Goals of breeding, aside from good harvest and taste: Ease of harvest, and resistance to cane blight, *Botrytis,* and the mosaic-carrying aphid *Amphorophora.*

Where to Plant

Raspberries are not undemanding. For good growth and bountiful harvests, they need a wind-protected, sunny spot with a fertile, medium soil. This should be deep, loose, and humus-filled, without standing water. Their optimal pH value is approximately 5.5–6, lightly acidic. If the pH is higher, iron and manganese deficiency may result. Light, sandy soils only are recommended with extensive additional watering, and fertilizing with humus-rich additives.

Late frosts generally do not bother raspberries; once in a while, a very early variety may be damaged. New shoots usually produce a somewhat later but nonetheless satisfying harvest.

Raspberries

Young raspberry stems: fully developed (*left*); sickly, not likely to survive (*right*).

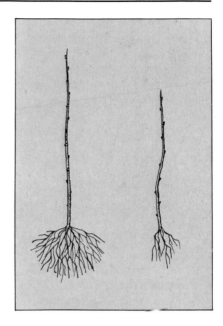

Planting

Like other bush fruit raspberries are planted in fall so the plants have a chance to root. If planting is done in spring, it should be as early as possible.

Distance: In the garden, 5 feet between rows and 15–20 inches between plants in a row is sufficient. Commercial growers may keep over 6 feet between rows —or, if using a *V*-configuration, 8–10 feet.

The roots of young raspberry plants are very sensitive and dry out easily. After digging them out from around the mother plant, replant immediately. Good soil preparation, as with the currants, is a given. Adding moistened peat or compost at planting time is important. But pay careful attention that the lower growth buds are not damaged when working around the plant—these are the canes that will bring next year's harvest. Plant the raspberries deeply enough, so that these buds are covered with an inch or so of soil. Water care-

Raspberries

fully after planting and mulch with peat or a similar material to maintain moisture.

Before planting, or even afterward, cut the canes back 12–18 inches. If they develop leaves, it is all right, because every leaf works for the plant; if not then it's still okay, because the buds around the roots are spared the energy and will grow that much more vigorously.

Shortened canes should not be allowed to carry fruit; instead, all of the plant's energy should be channeled into new growth. Very soon, older canes will die and can be cut back to the ground.

Raspberries love a protected place, and for that reason plant several short rows in a block rather than in a single long row. In the first two years, you can still use the room between the rows for vegetables, bushbeans, onions, leeks, kohlrabi, lettuce, or even strawberries—but only for two years. After that time the raspberries will need the room.

Pruning

Raspberries cannot stand alone, and after the second year they need supports. In the garden, plant them along a fence, but here there is always the danger of weeds which become a source of trouble in themselves. Freestanding plants need a simple wire support or, like blackberries, a V-shaped support.

Wire supports are simply posts, spaced a few yards apart, strung with two to four wires, the uppermost of which is about 6 feet high. It is also possible to run a second wire a few inches away so that the raspberry canes can grow up the middle, between the two. The additional wires can be wound around the same post, or attached to horizontal pieces that have been nailed into the posts. Double wires spare you the work of tying the canes to the wire.

The V-shaped support—not as easy to build—allows you to use both sides to bind on the canes produced in the previous year. There is room for nearly double the number of canes as with the single-wire support because plants can be set more closely together, and as a result the harvest is bigger. With this support, new young canes can grow undisturbed in the center of the V, and after the harvest the spent canes can be removed easily. Canes should

Raspberries

Raspberries belong in every garden. Here, two rows are growing on a wire fence.

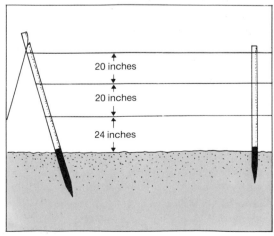

20 inches

20 inches

24 inches

Wire support for raspberries.

Raspberries

always be attached carefully to the wires with string, cord, plastic wire, or special clamps that hold them free of wire and prevent damage to their bark. Any bark damage may be an invitation to the dreaded cane blight.

Pruning raspberries is simple: Last year's canes provide this year's fruit, and they are cut to the ground or even deeper following harvest. In summer, sometime after June, or at the very latest in winter, new canes are thinned. For the first year after planting, no plant should be allowed to have more than two or three vigorous canes;

later, six to eight will be allowable. This way the planting never becomes too dense, and the remaining canes are much more robust.

Shoots growing between rows or too far away from the mother plant are removed. Otherwise, only canes that are weak, crooked, or those that are blighted are removed.

Finally, at winter's end, any very long canes may be shortened to 6 feet. Perhaps a few berries will be cut away at this stage, but the remaining ones will be that much better for it.

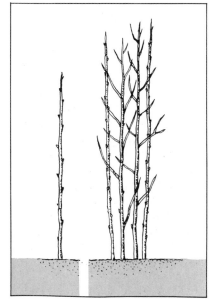

Pruning a raspberry cane (*left*).

Yearly removal of spent canes as soon as possible after the harvest (*right*).

68

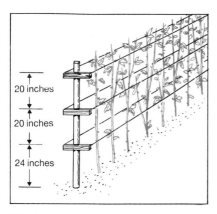

Double-wire support; raspberries grow in between.

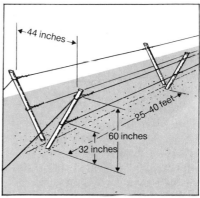

V-shaped support. Bearing canes grow to both sides, new canes grow in the center.

Soil and Fertilizers

The most important aspect of raspberry culture is providing enough humus for the plants. Raspberries originate in humus-filled soils and require abundant humus for their cultivation. Fine feeder roots in the upper soil surface require moisture that only a covered soil can provide. Mulching with manure or other organic materials, as well as maintaining a layer of humus on the soil surface from the time of planting onward, is of the utmost importance. Nothing else will do as much to protect the ground around the plant from drying sunshine as this kind of mulch. Naturally, it must be understood that any mechanical working of the soil—even the most super-

ficial—can damage plants, and this must be avoided.

Use organic fertilizers with raspberries because they are highly sensitive to mineral salts in the soil. Well-rotted manure, compost, enriched peat, bone meal, cottonseed meal, and other organic products are useful in maintaining a highly fertile soil. Raspberries are heavy feeders and will need moderate supplemental mineral fertilizing. 8 to 12 grams per square yard of pure nitrogen or 3 to 4 ounces of a complete fertilizer with 12 percent nitrogen per square yard. Raspberries are also sensitive to chloride salts; avoid using them, but choose a fertilizer with magnesium, because the plants quickly become deficient. Typical signs are the yellow

Raspberries

and necrotic areas between the veins on leaves. Anyone planning to use single fertilizers should choose slightly acidic materials. Raspberries require a slightly acidic soil with a pH of 5.5–6. If the pH is too low, lime with ground limestone or dolomitic lime. The best thing to do is to conduct periodic soil tests, keep a record, and remain well aware of the nutritional needs of the soil and its responses to fertilization and liming.

Lack of proper fertilization is a frequent cause of malnutrition in raspberries.

Diseases and Pests

Cane Blight
Dangerous and greatly feared in raspberries. Cause: The fungi *Leptosphaeria coniothyrium* and *Didymella applanata,* and sometimes *Botrytis cinerea* and *Elsinoe venata.* Infection enters through rips in the bark, scars on leaves, and similar damage, often through the damage of the raspberry cane fly *Thomasiana theobaldi.* Young canes show purple spots; the bark strips off and falls away. The canes grow but turn yellow early in fall, leaf out in the next spring, and die without blooming or bearing fruit. Bearing canes may also be stricken. Control is difficult: Destroy diseased canes. Mulch with manure, straw, peat, or similar materials to maintain even moisture to deter the condition. Spray for disease-carrying insects before bloom and with a fungicide at the end of May. Remove spent canes as soon as possible after harvesting.

Raspberry Beetle
The ¼-inch-long beetles *(Byturus tomentosus)* feed on buds and flowers. Maggotlike larvae feed on the fruit. Main damage: Raspberry maggots in the ripening fruit. Control: Spray after the flower buds appear and during bloom with an insecticide that is not harmful to

Raspberries

The most important fungal diseases of raspberries. The feared and widespread cane blight (*left, above* and *below*). *Botrytis* fruit rot (*above right*).

bees. Keep track of the period during which the poison is effective.

Botrytis, Fruit Rot
See the section on strawberries (page 108).

Virus Diseases
Of all the berry fruits, the raspberries, both wild and domesticated, are hit the hardest by virus diseases. 20 different viruses of raspberries are known; their contamination is due to both their general susceptibility to virus disease, and the wide range of different virus carriers. First among these is the big raspberry aphid, *Amphorophora rubi,* but cicadas and slugs do their part as well. Also to blame is continued vegetative propagation.

The symptoms of virus disease depend on the type of disease. Mosaiclike, chlorotic flecks along the main and secondary veins, and yellowing and banding of the veins are some of the most obvious

Raspberries

Leaf mosaic on raspberries.

signs of illness. Further results are limited growth, small harvests, fruit dropping before harvest, as well as small, often deformed fruit. Contamination with a virus is often fatal.

Control is difficult: Quickly dig out and destroy diseased plants. Use appropriate insecticides for fighting aphids. The first efforts to breed plants resistant to the raspberry aphid have begun in the United States, England, and in Germany. The best way to avoid problems is with a new planting of healthy raspberries propagated by meristem culture.

In the foreground: Stand of more or less virus-contaminated plants from a nursery. *Behind:* A healthy stand of meristem-cultured plants. Of the large number of raspberry virus diseases, the most widespread is leaf mosaic. It is the most dangerous, and it also occurs among blackberries.

Blackberries

The thornless blackberry.

Blackberries are fruit that for no good reason live in the shadow of others. For a long time they were planted as boundary fencing because their quick growth and wicked thorns made an impenetrable barrier. Berries were small, and the harvest, if anybody even bothered, was torture. And of course there was always trouble with the neighbors when the rambunctious plants overgrew their boundaries.

Actually the word "thorns," when applied to blackberries, is used incorrectly. Strictly speaking, blackberries have "spines." The difference is that thorns are a part of the wood, while spines are specialized epidermis. The immense number of spineless mutations and many spineless varieties are only a so-called chimera; only certain portions of the bark have been genetically altered—not the entire plant. For example, when a root is damaged, a new spiny shoot will sometimes arise from beneath the chimeric coating. Thus, when we speak of "thornless" varieties, please overlook the misnomer.

Thornless varieties were available right after World War II, making picking blackberries much more pleasant. Still, blackberries were not planted very frequently. And yet they have excellent culinary and health-promoting properties, and they deserve a larger following. In the marketplace, they are hard to find and expensive. Because it is so easy to have abundant blackberries at the peak of perfection, every gardener should have a stand.

Blackberries

Botany

The blackberry belongs to the genus *Rubus* and to the Rosaceae family. For a long time the botanical name for cultured blackberries was *Rubus fruticosus.* But listed in the Zander *Dictionary of Plant Names* (11th edition, 1979) under "blackberry" were 16 different species of *Rubus.* The earlier species name, *fruticosus,* isn't included anymore; instead, only the individual varieties of different *Rubus* species are listed. Of these, the most important are *Rubus discolor* (Theodor Reimers) and *Rubus laciniatus* (Thornless Evergreen). Therefore, instead of using the familiar *Rubus fruticosus,* we will use the designation *Rubus* species.

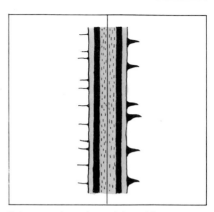

Spines as adaptations of the epidermis.

Blackberries, like raspberries, are biennials. Flowering begins in May, continuing well into August. The result is that, as is the case with raspberries, flowers and more-or-less ripe fruit are both found on the plant at the same time. Often fruit from the late flowers doesn't become fully ripe. All of the modern varieties are self-fruitful, with bumblebees and bees fertilizing the plants.

The berries are nutritionally sound, especially high in vitamin content—including all-important vitamin A (carotene) which blackberries contain more of than any other fruit. They are also extremely versatile and make excellent eating fresh, wonderful jams, compotes, jellies, and an exquisite juice.

The berries should be allowed to ripen fully; then they will not only have their complete nutritional content, but they also will have developed their marvelous flavor to perfection. When completely ripe, berries have a matte shimmer and detach easily from the stem.

Blackberries

Thorns: Extensions of the wood.

Propagation

The easiest way to propagate these rampant plants is simply to bury the tip of a branch in the ground at the end of August or the beginning of September. Dig a little hole, fasten the branch down with a wooden peg, and cover it with soil. Roots will develop in a matter of weeks, followed by a new shoot. Separate this new plant from its branch the following spring. Set it into a nursery row or plant it directly into the garden. This method of propagation by tip-rooting can be observed in nature with wild blackberries.

The upright growing species (like Wilson Fruehe) are the most easily grown by tip-rooting. Shoots arising from the roots around the mother plant may be dug up with a ball of roots in spring and transplanted as soon as possible; weaker ones should be given a year in a nursery bed.

Gardeners and nursery personnel, who deal in large numbers of young plants, and who have greenhouses, plastic tunneling, and sprinklers at their disposal, employ other methods of propaga-

In the United States, the motherland of blackberries, a large number of hybrids (crosses of blackberries with raspberries and other *Rubus* species) are grown that require warmer winters than those of northern Europe or the northern-most states. These include loganberries, boysenberries, youngberries, and dewberries—all popular where they are climatically suited. They are grouped together under the botanical name *Rubus loganobaccus*. Where they can grow, they should be because of their attractive appearance and excellent taste.

Blackberries

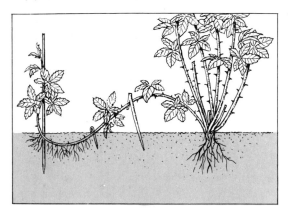

Tip-rooting blackberries.

tion. They can use green wood cuttings, short pieces of branch (two nodes each, up to forty cuttings per cane), and an old, dependable method, root cutting utilizing rooting hormone. There isn't room here to discuss all of the possible methods.

Varieties

Theodor Reimers Sand Blackberry. An American variety, very popular around the world. This extremely vigorous blackberry demands a great deal from its soil. But even in poor soil, its development is astounding: Rambling canes grow up to 40 feet long. Spines are numerous and sharp—even on the stems. Ripens from the beginning of August and continues for seven to nine weeks. Long fruiting clusters, berries medium to large, round, gleaming black, juicy when fully ripe, very sweet and aromatic. Good for eating and preserving. The wood is somewhat tender and should be laid down over winter and covered.

Black Satin. Bred in 1974 in the United States, a thornless variety. Growth is strong, with robust canes. Ripens two weeks earlier than most others, continuing to bear into October. Large, glistening black, juicy berries with a typical blackberry aroma. A very desirable variety.

Thornless Evergreen. One of the first spineless blackberries, available in the United States since 1926. Growth is moderate branching light.

Blackberries

Ripens about mid-August. Main harvest is somewhat later, but very large. Berries are large, roundly conical, firm, not too juicy, and finely aromatic like Theodor Reimers. Mainly for eating fresh. Wood is moderately hardy. Winter protection by laying and covering the canes is recommended.

The spineless blackberry Thornless Evergreen, with its prominent, deeply cut leaves (*Rubus laciniatus*).

Thornfree. Another American spineless hybrid, available since 1966. Growth is moderate. Ripens at the end of August. Berries are large to very large, conical, firm with a fine aroma. Good for eating fresh and preservation.

Smoothstem. Like Thornfree, spineless. Available in the United States since 1966. Growth is moderate. The variety has large to very large, round to conical fruits, that

This blackberry hedge is already in bloom.

are juicy but soft, only slightly aromatic and difficult to pick. Harvests are irregular. This variety is surpassed by others.

Wilsons Fruehe.   In many nurseries today, this is the only upright-growing variety available. Growth is moderate; 20 inches apart in a row suffices. Lightly spined. Ripens at the end of July. Berries are small to medium, long, oval, gleaming black, sweet, with many seeds.

Where to Plant

Blackberries can be grown in zones five through eight in the United States. In the northernmost regions or during uncommonly cold winter, canes can freeze (below 5 degrees F). Plants will then send out new canes but the following summer's harvest will be lost.

Laying the canes down and covering with evergreen branches, straw, or similar material will usually get the plants through the winter without damage. Blackberries are not recommended for altitudes over 1,500 feet because early mountain frosts occur before the harvest is over. The ideal spot for them is one that is in full sun, protected from wind and frost. They aren't fussy about soil, and in fact they can even be grown in sandy soils where no other kind of fruit will produce. The perfect plant for a tough spot is Theodor Reimers. On a wet, heavy soil, it bears so long and hard into fall, that a portion of the berries never ripen.

Naturally, it is to our advantage to provide the best, deepest, medium loose, humus-rich, and well-drained soil—especially when planting the more demanding upright varieties. But even when growing in a less than ideal soil, the use of a mulch to maintain moisture and fertilizing can work to promote healthy development and large harvest.

Blackberries

Planting

Because of the late ripening of the wood in autumn and to avoid loss because of winter weather, blackberries are usually planted in spring. It is best to use two-year-old plants with rootballs from a nursery. Distances between plants differ according to variety—whether upright or trailing. For the former, 5–7 feet between rows and 20 inches in the row is enough. With the latter, 10 feet between rows and 5–8 feet in the row is ideal. The further apart the plants are set in the row, the more canes per plant—up to a maximum of eight for strong-growing varieties.

When planting, take special care to avoid damaging both the weak roots and growth buds around the roots from which new growth will arise. Enriching the planting area with peat or compost is helpful. Plant the roots so that the buds are covered with a couple of inches of soil. Then water carefully and mulch the area with peat or similar material. After planting, the canes should be cut back to 8–12 inches.

Pruning

Pruning of upright-growing blackberries is similar to that of raspberries. This year's bearing canes were new last year, and this year's new shoots will bear next year's fruit. Spent canes should be cut down to the ground after harvest and burned.

But the strong-growing trailing varieties are a little harder to keep in order. While upright types can be tied to wire fencing, this isn't possible with trailing varieties. It is best to sink sturdy posts (4–5 inches across) 25–30 inches deep into the ground at intervals of 10 feet. Stretch wires at heights of 28 inches, 42 inches, 67 inches, and, possibly, 7 feet. To simplify distinguishing between the bearing and new canes of trailing types, a double-wire trellis with parallel wires or a *V*-shape support will make pruning and care easier.

Trailing blackberries require the same rotation of young and fruiting canes as the raspberries or upright varieties. The trailing canes can be espaliered in a fan or palm shape, which is suitable for the less robust types, or, for varieties with very long, trailing canes, simply woven or looped along the wire trellising. How they are trained is a matter of taste and convenience.

Blackberries

Blackberries on a *V*-shaped support.

Methods of training blackberries. Fan-shaped espaliered canes (*above*), palm-shaped method (*center*), weaving system (*below*).

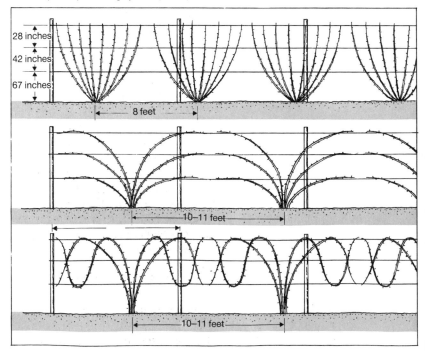

Blackberries

Summer Pruning

Most important is the summer pruning. Allow four to eight strong, young shoots to emerge from the ground, cutting back all others. Every two or three weeks tie the young canes onto the trellising. If you keep this up, it will be easy to distinguish between canes and the planting won't turn into a wilderness. Around the end of June, the new growth will develop abundant side growth that will become 10 feet long if it isn't shortened. Cutting back this new growth to two to four buds as soon as it reaches 1 to 1½ feet has to be done several times before fall. It is the only way to maintain order in the planting.

Early in summer, the new branches will again send out shoots that have to be shortened again; laterals may be used instead of young canes if there are not enough. If summer pruning is neglected, all the canes will be covered with flowers but they will develop only a few small berries. In addition, it will be a miserable job to harvest the fruit. Prune as directed and the vigorous basal nodes will provide abundant blooms and big, perfect berries that are quick and easy to pick.

Thus the most important pruning is done in the summer. In late winter, you'll have to cut out any frost-damaged growth and cut back by a third canes that have become too long—Theodor Reimers can grow up to 40 feet. In any case, the tips of canes are less well developed and would only produce inferior fruit. This cutting back boosts the flowers and fruits on the remaining portions of the cane. If you have neglected to cut back and tie on the new growth in summer, you can still do it in winter—more drastically. Cut the side shoots back to one or two buds.

Soil and Fertilizers

Humus is just as important for blackberries as it is for raspberries—especially in sandy soil.

Blackberries

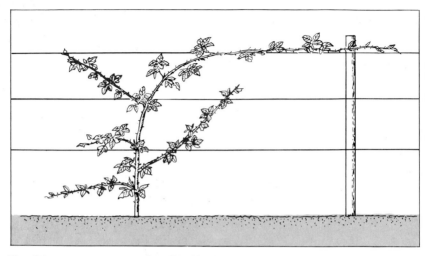

The all-important summer pruning of blackberries.

The ground around plants should be mulched with an organic material like manure, peat, or straw. It may be the same material used to protect the plant from winter cold. Although mulches help to keep the soil moist, during periods of drought in the summer, supplemental watering may be necessary. Fertilize the same as with currants. In the event one doesn't use mulch, great value is placed on additions of organic material, well-rotted manure, peat, bone meal, and so on, to the soil, and chemical fertilizers are also used. Liming should be done (up to pH 6.5) as with currants. Chemical fertilizers should be divided into thirds—two-thirds applied in spring, the remaining third after blooming.

Diseases and Pests

Blackberry virus diseases: Leaf blotch, stem-end rot, red-leaf spot. The causes are the *Rhabdospora ramealis* and *Gnomonia rubi* fungi. The former strikes in April to June; the latter later in summer. Red to purplish brown spots appear on the canes, often encircling them. Diseased growth withers and dies. To control: For *Gnomonia,* there is no known remedy yet; for *Rhabdospora,* a prophylactic copper-based spray may be useful (spray at the end of April to mid-June).

Blackberry Mites
Blackberry mites, *Aceria essigi* (often called *Eriophyes essigi*), can do enormous damage to fruit: They suck juices from the half-ripe berries, which become prematurely red and remain that color, becoming hard, sour, and inedible. To control: Spray mid-May and during the blooming period with a sulphur-based organic fungicide.

Botrytis, Fruit Rot
See the section on strawberries, page 108.

Blackberry virus disease (*above* and *center*). *Botrytis,* fruit rot (*below*).

Strawberries

If there is a "queen of berries," it has to be the strawberry. Of incomparable flavor, strawberries produce abundant harvests in no time at all, and even the smallest garden has room for them. A wide range of fruiting times allows us to enjoy this excellent fruit throughout the summer season, into fall.

Strawberries are not only a highly prized fresh fruit—eaten, perhaps, with sugar and cream —but useful for every imaginable sort of preserving or cooking purpose. Compotes, jams, ice cream, tarts—even juice and wine can be made from them, and much more. They also freeze well for winter use.

Botany

The strawberry belongs to the genus *Fragaria* and the family Rosaceae. The big, beautiful berries of our gardens are grouped under the name *Fragaria ananassa.* At the beginning of the eighteenth century this species was obtained from repeated crosses with the wild North American species, *Fragaria virginiana,* and the South American *Fragaria chiloensis.* Other kinds include the wild European strawberry, *Fragaria vesca,* from which the Alpine strawberry, *F. vesca* (var.) *semperflorens* was developed. Alpine strawberries are propagated by seed. Botanically speaking, a strawberry plant is a perennial, whose roots survive winters underground. Strawberry crowns are studded with low-growing foliage. Flower stalks and runners at whose nodes new young plants develop grow from the leaf axis. The flowers of modern strawberries are bisexual: They possess both masculine and feminine reproductive organs. Earlier varieties like Mieze Schindler, Direktor Paul Wallbaum, and Macherauchs Spaeternte had undeveloped or totally absent stamens, and these purely female plants always needed other varieties in the vicinity for pollination, remaining barren when grown

Strawberries

alone. Today these varieties are no longer found.

Even when the flower of a given variety has a stamen, it can happen that the pistil of the same flower is physiologically incapable of self-fertilization, so that the variety—just like the purely female plant—needs another variety for fertilization to occur. For safety's sake, it is always better to plant not only one variety, but several.

A strawberry is an aggregate fruit, the product of separate ovaries within a flower uniting to produce a single fruit; around a fleshy, enlarged torus the delicious strawberry "fruits" are clustered like little nuts. Quantity, arrangement, and function of the ovaries are decisive factors in the ultimate size and shape of the berries.

Wind, insects, and above all bees fertilize the flowers. Inadequate fertilization after cold, rainy weather, or because of a low bee population, may result in somewhat malformed fruit. The more ovaries that are fertilized, the more beautiful and perfect (through hormonal stimulation) the fruit will be.

Starting New Plants

Growing new plants couldn't be easier. Plants are allowed to send out runners that eventually root; when these runners are developed enough, they are carefully transplanted. The gardener can help the process along by loosening the soil between rows and enriching it with compost or moistened peat so that the new plants will be able to root more easily than they would in soil that is compacted by foot traffic. Another method is to prepare planting holes between rows by filling with peat or compost. When the strawberries send out runners, these can be anchored into the already prepared holes. This method encourages the development of a little rootball to facilitate transplanting. Runner plants that are weak can be transplanted into a well-prepared nursery bed until fall or the following spring.

It is worthwhile to keep in mind the mechanics of runner production. If flowers are not removed and a full harvest is allowed to develop, the number and vigor of runners produced is diminished. And again, soil compacted by foot traffic during harvesting makes it extremely difficult for new plants to root.

There is yet another danger. It is well known that strawberries, if not continuously and systematically renewed, diminish in yield and vigor over the years. People speak of the plants "giving out" without really knowing what causes it. Today we know that giving out is almost always the result of disease

Strawberries

Propagation of strawberries by tissue culture "in vitro."

or pests like slugs, mites, fungi, viruses, and the like. A stricken plant may show no signs of affliction for a while, becoming noticeable only after the plant is severely diseased. Whether or not a runner plant will also be afflicted is nearly impossible to predict.

It is imperative to introduce new plants into the planting from time to time; these can be purchased from one of the many hundreds of specialty nurseries that sell certified, healthy stock. There, only the most promising, high-yielding plants are propagated and the latest cultivation techniques employed. The variety and health of the young plants are guaranteed and certified to be virus free. These firms frequently offer a wide selection of plants to suit different purposes and conditions; choosing

the most suitable from among the many available ensures that the gardener will enjoy continued high yields and healthy plants.

Every gardener should make an effort to grow the very best plants available, and one only has to choose and buy them from a reputable source to benefit directly from the enormous amount of research and the years of hybridizing that has been carried out around the country. For the home gardener, the main interest should not be for the fruits commonly grown commercially (which may or may not be chosen for their qualities endurance and appearance), but for those with the best taste and the newest hybrids with the highest resistance to disease.

Alpine strawberries are not propagated by runners, but by seed—

These fruits have been tied up to avoid becoming soiled on the ground.

and, infrequently, by division. One can also grow the big, runner-type strawberries from seed, but the result will be a colorful assortment of different variations of the mother plant. Rarely will superior individuals be grown from seed. Strawberry seeds appear tiny and half-buried on the skin of the fruit. To collect them, allow a fully ripe berry to dry on paper. Either the little seeds will fall out by themselves or they will be easily loosened from the dried skin. Sow them immediately or in the following spring in flats on a peat-sand mixture. As soon as the seedlings have three to five leaves, they should be separated, and when strong enough planted in the garden.

A relatively modern method for propagating strawberries is tissue culture. The first work in this field was undertaken in the United States in 1962 to eliminate a virus disease. In Germany, R. Hummel of Stuttgart/Weilimdorf was one of the first to use tissue culture with strawberries. With this method, the meristem, a minute portion of the undefined growth tissue, is collected and grown into a little plant on a sterile, suitable medium "in vitro," and vast numbers of young plants can be obtained this way. These plants are absolutely free of fungi and bacterial pests, and almost always free of viruses.

Varieties

Ever since the first plants worthy of culture were grown in England two centuries ago, thousands of others have been bred and the development of still more is currently underway. There is so much activity in the realm of breeding new varieties that it is difficult for even an expert to keep abreast of all developments. But only a very few varieties have withstood the test of time: Many new hybrids are praised to the skies today, only to be gone and forgotten tomorrow. There is no list of plant varieties that ages as quickly as that for strawberries.

Choosing from among the varieties is not simple: Gardeners on the one side and commercial growers on the other—each have to come to terms with different qualities. For commercial purposes the following attributes are taken into consideration: ripening time, quantity of harvest, ease of picking, appearance, color, texture and size of fruit, resistance to mildew, diseases, pests, ability to withstand transportation, and suitability for cooking. The home grower has fewer demands. The gardener puts great value on a good harvest, appealing aroma, health, and versatility of the fruits—especially their freezability. Because the gar-

dener has nothing to lose, he can experiment with a few varieties.

Choosing a variety is complicated by environmental factors. Climate, soil, latitude—all have an effect on the performance of any given variety, as expressed in quantity of harvest, resistance to disease and pests, and the degree to which the plant adheres to the archetype of the variety.

The table on pages 92–93 includes only varieties that have proven themselves and are fundamentally worth recommending; it is impossible here to thoroughly describe all plants and fruits. Ripening times are provided to allow for the selection of plants to ensure a long season, vigor, to determine distance between plants, quantity of harvest, type of soil, and a few other distinguishing characteristics of a variety —among them resistance to diseases and pests.

New and tried-and-true strawberries: Senga Precosana (*above*), Senga Pantagruella (*center*), and Senga Sengana (*below*).

Strawberries

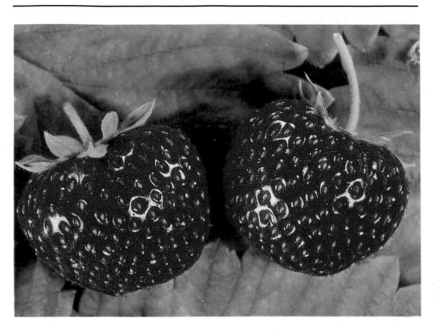

Senga Litessa, a mid-season, strong-growing, firm strawberry variety.

Naturally, this assortment is incomplete: It includes a variety or two that no longer dominate the commercial field, but can be safely grown by the gardener without disappointment—perhaps with pleasure. The list of the big berries has grown through the addition of new hybrids. The following are named:

Early: Deutsch-Evern's Fruehe (PDE 101), Deutsch-Evern's Finessa, Karina, Pocahontas, Macherauchs Fruehernte.

Mid-season: Elvira, Senga Dulcita, Senga Fructarina, Senga Gourmella, Senga Litessa,

Strawberries

Important Varieties of Strawberries

Variety	Ripens	Growth	Harvest
Senga Precosa	very early	moderate	medium to small
Senga Pantagruella	very early	moderate	high
Zefyr	very early to early	strong	medium
Regina	early	moderate	medium to small
Gorella	early to mid-season	moderate to strong	medium to high
Senga Precosana	early to mid-season	strong	high
Macherauchs Marieva	early to mid-season	strong	medium
Hummi-Grande	early to mid-season	strong, very long runners	medium
Senga Gigana	early to mid-season	very strong, strong runners	medium to high
Vigerla	early to mid-season	moderate	high to very high
Georg Soltwedel	mid-season	moderate	medium
Senga Litessa	mid-season	very strong	very high
Senga Sengana	mid-season	very strong, strong runners	high to very high
Hummi-Ferma	mid-season	moderate to strong	high
Silvetta	mid-season	strong	high
Tenira	mid-season to late	moderate to strong	high
Red Gauntlet	mid-season to late	strong, early, many runners	high to very high
Asieta	late	moderate	medium
Deutsch-Evern's Solweta	late to very late	very strong	high to very high
Elista	late	moderate	high

Strawberries

Remarks
C = commercial use   H = home garden

Needs highly fertile soil, *Verticillium* wilt and red spider, some *Botrytis*, C + H

Botrytis, H

Mildew, C + H

Mildew, *Verticillium* wilt, *Botrytis*, H

Low soil fertility, large fruit, little aroma, *Verticillium* wilt, C

Moderate soil fertility, *Verticillium* wilt, some *Botrytis*, C + H

Moderate soil fertility, mildew, *Botrytis*, H

Giant, irregular fruit, not aromatic, scarcely any *Botrytis*, C

Low soil fertility, very large, aromatic fruit, *Vericillium* wilt, very susceptible to *Botrytis*, C

High soil fertility, mildew, little *Botrytis*, H

Moderate soil fertility, mildew, red spiders, moderate *Botrytis*, H

Firm flesh, *Botrytis*, C + H

Best-known German variety, important in industry, low fertility, large, firm, highly aromatic fruit, very good for freezing, susceptible to *Botrytis*, C + H

Firm flesh, robust

Fruits large to very large, little aroma, *Botrytis*, *Verticillium* wilt

Fruit large, firm, good aroma, slight weakness to *Botrytis* and *Verticillium* wilt, red spiders

Low fertility, large fruit, no aroma, no *Botrytis*, C

High fertility. *Verticillium* wilt, little *Botrytis*, H

Very large fruit, strong runners, *Botrytis*, C + H

Soft flesh, little *Botrytis*, C + H

Strawberries

Deutsch-Everns's Frikonsa, Hummi-Stugarta, Induka, Karona, Splendida, Deutsch-Evern's Bromba.

Late: Hummi-Ferma, Bogota, Domanil, Deutsch-Evern's Famosa, Tago, Talisman, Siletz, Direktor Paul Wallbaum, and many more.

Everbearing Varieties

In addition to the large group of strawberries that bear once each year, there are those which bear twice or more, the everbearing types. Recommended are the very early Hummi-Eroma as well as the varieties Hummi-Gento and

Ostara. Planting these kinds is only for gardeners who are ready to put a little effort and elbow grease into the planting. The first harvest arrives in June, about half the size of the single-bearing types. As soon as it ends, blossoms appear for the second and main harvest. Only well-nourished plants that have been fertilized and watered are capable of this effort. Although single-bearing varieties recoup their energies in the period after harvest, developing next year's buds, the everbearing kinds may exhaust themselves through the late harvest. To prevent the plant's going into winter in a weakened condition—they are just as hardy as the single-bearers—it is advisable to cut off any buds that appear after the beginning of October. With good fertilizing, the plants can build up a little reserve for the coming winter. Then, if late frosts destroy the single-bearer's crop, the everbearing varieties will make sure the gardener does not want for strawberries.

One new development for the home gardener is the climbing strawberry Hummi. The first harvest arrives about the time that early single-bearing varieties ripen. As soon as the late single-bearers finish, Hummi's second harvest is underway. Simultaneously numerous runners develop—all of which have to be tied onto fences, cords,

The late strawberry Domanil.

94

Hummi-Stugarta.

The climbing strawberry Hummi.

Strawberries

Alpine strawberries in windowboxes for the balcony or terrace.

Alpine strawberries bear throughout the summer.

or poles because this variety needs help to climb. Along these runners, which can grow to 4 or 5 feet, many small rosettes bloom and fruit develops. By August, Hummi is bearing fruit on newly developed runners. Berries are medium to large, and very aromatic. The harvest is quite large because fruit is borne constantly from early summer until frost. This variety will provide two good years of harvests.

Alpine Strawberries

Finally, a word about Alpine strawberries. One rarely sees these non-vining woodland berries in the garden anymore, even though that is the best place for them. Certainly their berries are small, but excep-

tionally aromatic and perfect for the preparation of strawberry punch. The best-known varieties are the old ones Ruegen and Baron Solemacher, as well as Rimona-Hummi (bred in 1978), with fruit in July, August, and September. Because of their extraordinary growth, the latter variety is especially good for planting in April. It will bear a full harvest the same year. These small plants are exceptionally well suited for lining paths or for planting in bowls, pots, and boxes on balconies.

Strawberries

Where to Plant

Our climate supports the growth of strawberries in most parts of the country. But different climatic conditions will of course affect the plants' performance. When vegetative growth begins in spring, the length of the growing season, temperature, wind, and water are factors of the microclimate, which (especially for the commercial grower) is of great importance in choosing which varieties to grow and how to grow them. The most important consideration in installing a strawberry planting is the danger of late frosts. Low-lying frost pockets with no place for the cold air to go must be avoided. Late frosts—temperatures under 32 degrees F—during flowering will damage the ovaries and turn the flowers black. Straw covers provide some protection and may be used after frost danger has passed to lay beneath fruiting plants.

An enclosed place, without air circulation, is not a good spot for strawberries. Dew and rain will dry up slowly and *Botrytis* may strike, causing rotting of the fruit. The opposite situation, a too-open, windy place, can ensure that plants—whether in summer heat or winter frosts—always suffer from lack of water. A nice sunny place in the garden, possibly half-

Late frost damage (flower on *right*) in strawberries.

shaded, with some wind protection from trees and shrubs, is the perfect place for a strawberry bed.

Strawberries will tolerate a wide range of soils. Humus-rich loam, loamy sand, or sandy loam as well as pure humus are all excellent. But even a light, sandy soil will work if it is enriched with humus, fertilized, and watered. Sandy soils are warm and may speed up the harvest. The soil should be slightly acid to neutral—but in this respect, too, there is a great deal of leeway.

Planting

Before planting a deep working of the soil is recommended. Strawberries penetrate the soil as deeply as the bush fruits (up to a yard

Strawberries

Robust young plants.

deep) and are not the surface-rooted plants that they have long been considered. There are a number of possible planting times; the best and most common is the summer planting at the end of July or the beginning of August. The earlier planting is done, the larger will be the next year's harvest. Planting in the middle of September will provide barely half the harvest of a planting undertaken at the end of July or the beginning of August. If the beds aren't free in August, plants can be heeled into good, fresh soil, and transplanted toward the end of September with the biggest possible rootball.

The preference for summer planting has physiological reasons. Roots develop most intensively in August and virtually stop their development at the end of Sep-

The right planting depth will determine whether or not a plant thrives.

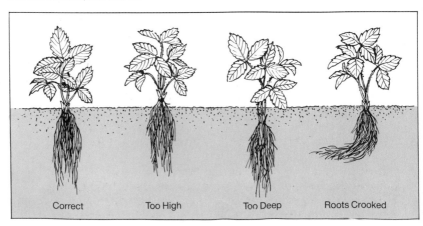

| Correct | Too High | Too Deep | Roots Crooked |

Plants from cold storage: Senga Sengana.

tember. What follows is a speedy development of young plants, and the stimulation of flower buds through the shorter days and dropping temperatures.

Finally, robust young plants store carbohydrates before winter dormancy as reserve energy for next spring's growth.

Anyone considering growing strawberries on an annual basis should select July for planting; it is the only way to obtain a good harvest the following year. The grower should ensure that plants are adequately watered during dry weather, so they may develop without interruption. With later planting, two years should be allowed the first harvest will be minimal while the second-year harvest will be especially large, although the berries themselves may be slightly smaller.

Another possibility for planting is spring—March or April—as soon as the first leaves show on the little plants. In such a case, there should be a modest harvest the first year. But if not planted until late April, it is better to save the plant's energy

and cut off the blossoms as soon as they appear; the following year's harvest will more than repay the loss.

The third deadline for planting strawberries is only of use to commercial growers: mid-June to mid-July. At this time strawberries which have been refrigerated can be planted. Fully developed, these are young plants that were removed from the ground during their dormancy and stored at temperatures below freezing. All leaves are removed and the plants are packed in airtight plastic. In

Strawberries

order to obtain a successful harvest, planting needs to be done in early June to equal that of fresh green plants set in the ground in July–August. These frozen plants also demand a good bit of work after planting because their flowers and runners must be removed. For a home garden, they are not practical.

Distance between plants is determined by the type of planting, the length of planting (one or two years), the individual plant's vigor, and the fertility of the soil. One-year plantings require 12–16 inches between rows for moderately growing plants and 16–20 inches for the more vigorous ones. Within the row, distances can range from 6–12 inches apart depending upon the rate of growth. If strawberries are planted in beds, the customary 4 feet in width, there are a number of planting possibilities: three rows, 16 inches apart, is just one option. If the plants are especially expensive or hard to find, they can be grown in two rows a good distance from each other. In the second year, the rows can be filled out with runner plants, and a third row started. This would not be possible, however, with three rows at 16 inches apart.

Another variation is band planting. 2½ feet is left between double rows or bands, each 6 inches apart. Large commercial plantings will allow 2½–3 feet between rows and 12–16 inches between plants to allow for mechanical tillers and similar equipment.

Never plant too closely together: The sun has to be able to hit the ground in order to warm it, and air needs to circulate around plants in order to prevent *Botrytis* fruit rot.

Strawberries

The ground between two rows of strawberries has been covered with plastic mulch.

Growing strawberries with black plastic mulch.

Black Plastic Mulch

Every once in a while commercial growers or gardeners may be observed covering beds with black plastic. What does it do, and what are its advantages and disadvantages?

Before laying on the plastic mulch the same thorough preparation of the soil employed with other planting methods is in order. In addition, for a one- to two-year bed, fertilizer has to be added at the rate of 20–25 pounds per 100 square yards—and worked in at least 8 inches deep. Shortly before planting at the end of July or the beginning of August, strong black plastic mulch should be rolled out over the moistened ground and weighted down with earth along the sides. "X's" are cut and plants inserted. Some known advantages of this method include the maintenance of good tilth below the covering, the virtual absence of

Strawberries

weeds, obviating the need for hoeing, and the early warming of the ground to speed up the harvest. In addition, straw covering to keep fruit clean is no longer necessary.

Disadvantages include the cost of the plastic, the additional work of laying it out and removing it, and the increased likelihood of frost damage because the ground stays warmer longer. In addition, rain falls on top of the plastic and evaporates without penetrating the soil. Too, very light soils can dry out quickly during heat waves. Finally, new plants cannot be started using plastic mulch. The pros and cons of using black plastic mulch are about equal. It is up to the gardener to experiment.

Peat Walls

Peat walls are an interesting and novel way of planting strawberries that appeals to the handyman in all of us. A freestanding framework filled with peat is constructed and planted (see illustration). This is the ideal way to grow strawberries on a terrace or deck.

Life of a Planting

The question of how frequently to replant is easy for the commercial grower. If one plants at the end of July or the beginning of August, to let the young plants develop fully, one can count on large harvests of big berries the following year.

The planting can pay for itself in a single year and be replaced thereafter. Often, growers allow the planting one more year because

Strawberries growing in a freestanding, peat-filled framework.

Strawberries

Straw mulch under strawberries is the best defense against *Botrytis* fruit rot.

Straw mulch. Traffic boards protect the ground from compaction.

the second harvest is almost always more abundant in yield but composed of smaller-sized fruits. Picking costs are higher the second year, but can be offset if plants are given over to self-picking, a popular alternative. Two years is the maximum life of a commercial planting.

It is often stated that in a home garden good care, fertilizing, and watering can make a strawberry bed last five years and longer. This was truer formerly than it is today. The reason for a faster replacement is that over the years, along with increasing fruit size, there has been a steady rise in the prevalence of strawberry mites. For this reason, home gardeners, too, prefer one- or two-year cultures.

Soil

If it isn't mulched, the ground between the rows will have to be hoed from time to time to keep it from crusting, drying out, and becoming choked with weeds. There is a saying, "Once hoed is twice watered." Nevertheless, even with regular hoeing, you'll have to water during dry spells.

If the ground is mulched with an organic material such as straw, peat, or something similar, one doesn't have to fear that the ground will dry out. Under such a mulch, the soil remains fresh and moist. In addition, the fruits are shielded against mud and can be picked cleanly and easily (an important advantage).

Strawberries

Fertilizers

The most important fertilizer for strawberries is humus added to the soil. If there is abundant humus in the soil, supplemental mineral fertilizing is often superfluous with no effect whatsoever on the plants. Humus is best added before planting at the rate of 100 pounds per 100 square yards of well-rotted manure, or 4 bales per 100 square yards of peat, or by preplanting the area with lupines or other plants that improve the soil.

Mineral fertilizers can be added a few weeks before planting as well at the rate of 5 pounds of chloride-free, complete fertilizer per 100 square yards. After five or six weeks, when the plants have clearly grown and are leafing out, an additional 5 pounds may be added.

Fertilizing will encourage the production of robust young plants during the fall of the initial planting year. Use the same dose the next year immediately after harvesting, when the rows have been cleaned up and the soil loosened. Repeat four to six weeks later. In spring, adding nitrogen is done only if the plants look weak and clearly need nourishment. Otherwise, too much nitrogen given in spring fosters only excessive foliage growth and *Botrytis*.

It is possible to use single fertiliz-

Deficiency of trace elements in strawberries: chlorosis due to iron deficiency (*above*); manganese deficiency (*below*).

ers in place of the complete types, but it is more practical only when working with very large plantings. The following quantities are for use on 100 square yards: 4 pounds of ammonia sulfate or sodium nitrate; 4 pounds of bone meal or super-phosphate; and 7 pounds of muri-ate of potash. These quantities should be divided up and applied in the same manner as the com-plete fertilizer. All of these fertilizers shouldn't touch the leaves of the strawberry plants but be strewn between the rows. Watering after fertilizing cleans off bits of fertilizer from the plants and dissolves the fertilizer.

Further Care

Further care of a strawberry plant-ing includes protection against cold. Especially in areas of hard winters, strawberry beds should be covered with evergreen boughs or straw. The plants can withstand some frost and should be uncov-ered in spring as soon as possible.

More dangerous and fearsome than winter frosts are late-spring frosts when the plant is in flower. Commercial growers protect against these late frosts with artifi-cial rain. In the garden, cover the bed completely and fairly thickly with straw. It will protect against

light frosts, and won't harm the plants if left on for several days.

Another measure that is recom-mended is the removal of all runners. Of course, no new plants will be produced but the mother plant will be that much stronger for having only to produce fruit—instead of both fruit and new run-ner plants.

Once in a while, you'll hear about mowing down the foliage after har-vest. The value of this practice is disputed. Nobody is really sure whether removing the foliage weakens plants, something which will be expressed later in a dimin-ishment of the harvest. Mowing the leaves away makes sense if the planting is suffering from a fungal disease like mildew or white spot, or is infested with mites. Mowing is no substitute for other means of controlling these problems, but may possibly help out. When a plant is infested with mites, spray-ing the crown of the plant is espe-cially important. Mowing will make the crown more accessible. Foliage that has been mowed off should be carefully gathered up and burned.

Strawberries

Diseases and Pests

Strawberry Mildew

The botanical name for the carrier of strawberry mildew, *Spaerotheca humuli* (*humulus* is Latin for hops), indicates that hops are one of the host plants of this fungus, of which a particular race has adapted itself to strawberries. For a long time, this disease affected only early ripening types, but today it has spread to many mid-season and late varieties. Plants most at risk are those in closed-off areas on quickly drying, humus-poor soils, which don't allow for quick, undisturbed growth.

Humid, warm weather, and an excess of nitrogen, are other factors that likewise foster mildew.

Sphaerotheca humuli belongs to the true mildew fungi, whose mycelia live on the surface of the plant but which send appendages into the tissue for nourishment. Most often the fungus occupies the underside of the leaves, building a fine, white, floury coating in which masses of spores are produced that spread the infection.

The edges of the leaves roll inward; the undersides of the leaves exhibit a reddish color. The fungus can cover the stems and even the flowers, whose petals turn pink and cease normal development. When fruit is infected it stops growing, berries stay small, become hard, dry, and inedible. Usually ripe fruit is unaffected because the disease often gains momentum only after the harvest.

For this reason fighting the fungus is usually worthless. Regularly repeated sprays with the usual fungicides are possible, but it is far better to replant with a variety that is resistant to mildew.

Strawberry diseases and pests. *From above left to below right*: strawberry mildew, leaf blotch, grey mold (*Botrytis* fruit rot), nematodes, strawberry mites, strawberry beetle, snails/slugs.

Strawberries

Leaf Blotch

This disease, caused by the fungus *Mycosphaerella fragariae,* is more noticeable than it is dangerous. After harvest in summer, round white flecks with brown or red rims appear on the leaves. Often the center falls away. In wet weather, the flecks run together, the leaves dry up and fall off. To control, collect the dead leaves in winter and burn them. For very serious cases only, spray after harvesting with a copper or Zineb preparation.

Botrytis, Fruit Rot (Gray Mold)

In wet years and moist places this is a serious disease. Only a wine-grower might welcome this organism for the special and prized Edelfaele harvest. A serious problem with strawberries, raspberries, and blackberries, *Botrytis* is caused by the fungus *Botrytis cinerea.* The fungus winters on dead plant material, producing masses of spores in the spring. These germinate on the flower parts and grow into the developing fruit tissues. Often green fruit will already exhibit rotten spots. On soft fruits, a gray velvety coating appears, especially in wet weather, composed of the spore cases that have opened to release countless other spores.

To control: Plant in a well-aerated place. Don't apply nitrogen late in the season. Place straw under the developing berries.

Plants should dry off quickly after dew or rainfall. Remove dead-plant material immediately. In wet years use special fungicides. Spray two or three times during bloom, at eight-day intervals. Raspberries and blackberries need several sprays during the blooming period as well: Apply the first spray when 10 to 20 percent of the blossoms are open.

Nematodes

Nematodes are tiny worms, only visible through a microscope, that live either as ectoparasites on the surface of young leaves or buds, or that burrow into the leaf axis and buds to bore out cells and drain them. The first infestation comes from the soil in which nematodes can live. With strawberries, the main types of nematodes causing disease are known: The strawberry or leaf nematode (*Aphelenchoides fragariae* and *A. ritzemabosi*), and the stalk nematode (*Ditylenchus dipsaci*). Often plants are infested with both.

Symptoms of disease are a general weakening of the plant in the beginning, diminished yield, and only a few runners. The term "giving out" is used without knowing the exact cause. In bad cases, clearly recognizable in spring, the leaves turn a gray-green, stay small, and appear more or less deformed or crippled—stiff, brittle, often reddish. The stems stay short and are thickened and wrinkled. In severe cases, the flowers are damaged as well. Flower stalks are short and broadened; the flowers themselves are completely deformed, green, and often clumped together. Sometimes this clearly abnormal appearance is called the "cauliflower sickness." Besides the nematodes, there is probably also a bacterium at work, *Corynebacterium fascians.* The few fruits that do develop are small and deformed. From June on, new, normal-looking leaves grow and the signs of the illness are more or less hidden. Runner plants can be infected even though they exhibit no symptoms.

Controlling nematodes directly with sprays or systemic preparations is not possible. Infected plants should be dug out and burned (not thrown on the compost heap). Prevention is the best cure: The gardener should make sure that the plants he buys are healthy. He can leave the production of healthy, nematode-free plants through chemical soaking of the soil, warm-water treatments, and fruit exchanges with the grower.

Strawberry Mites

The strawberry mite (*Tarsonemus pallidus*), a soft mite, is one of the most dreaded pests in strawberry cultivation—it causes enormous damage. Unlike the red spider mite, which appears in dry conditions, the strawberry mites occur in moist, warm conditions. The young crown of an afflicted plant is frilled, stays small, sickens, and dies. Flowers and fruits are deformed. Masses of dirty-white mites are visible on the crown when viewed through a magnifying glass. Up to seven generations can be seen. Important: Use healthy plants from a good nursery (where mites can be controlled by gassing with methylbromide). Use annual cultivation. If necessary, mow plants and spray the new growth thoroughly twice with endosulfan. Make sure that the crown is well sprayed.

Strawberry Beetles

Small (under ¼ inch), black beetles, *Anthonomus rubi,* appear when fruit ripens. Eggs are laid from May to June in the still-green flower buds. The stems of flowers are bitten; they flop over, and they wilt. Inside buds are dirty-white larvae. To control: Dust or spray before bloom with insecticide (take care not to harm bees). Raspberries and blackberries are also attacked.

Slugs, Snails

Slugs of different types eat the ripening fruit. Frequently they carry

Strawberries

Botrytis. In the daylight only the slimy trails of slugs are visible. To control, catch slugs in traps made of boards, flowerpots, and so on. Surround the strawberry bed with a broad band of dry pine needles a couple of inches thick, or lime, ashes, etc. Snail pellets, generally containing metaldehyde, are sold in garden shops.

Virus Diseases

Viruses present a serious problem to the commercial grower; the most virulent cause yellowing of the leaf edges, the ruffling sickness (both of which have been known since 1974), as well as the Arabis mosaic. Other symptoms are small or deformed leaves, premature fall color, retarded growth, and diminishing harvest. Aphids carry viruses, especially *Pentatrichopus fragaefolii*. Preventative control: Spraying against aphids; destroy infected plants; use only plant material propagated by meristem culture.

Strawberry Breeding

Spadeka is a new variety with the aroma of wild strawberries.

Of exceptional interest in the realm of strawberry culture is a new hybrid developed by the Max-Planck Institute for Hybrid Research in 1972 and patented in 1974. After many attempts, crosses between the garden berry *Fragaria ananassa* and the wild strawberry closest to the Alpine, *Fragaria vesca semperflorens,* were achieved. Of thousands of these hybrids, the Spadeka possesses excellent fruit quality, the aroma of wild strawberries, as well as robust growth and the best potential for a "strawberry meadow."

In appearance individuals of the species *Fragaria ananassa* and *F. vescana* hybrids don't differ as much as they do in growth and productivity. The hybrid sends out runners late into the fall, even before the mother plant is fully developed. In the spring the plants begin to form a strawberry meadow—a term for a new mode of growth that is reminiscent of wild strawberries. After the plants form a solid mat, they send out runners only on the edges of their meadow. They then grow into a steady state in which production of leaves and fruit is somewhat diminished, but continues for some years. For commercial growers using tradi-

Strawberries

A strawberry meadow of the new hybrid Spadeka.

tional row culture, Spadeka's yields are insufficient, and the fruit does not travel well.

Runners continue to develop until tall. When the plant turns its energies to the production of flowers, they are carried not on the usual small stems but on a single, robust stalk with seven to twelve bisexual flowers, held (as the fruit will be later) at the same level as the leaves, about a foot above the ground. Fruits are rarely soiled; they dry off swiftly after a rain and are well protected from *Botrytis* fruit rot.

Berries are early to mid-season; they are medium to small, round to kidney-shaped, shiny, medium red, soft and juicy, with a firm, elastic skin. Scent and taste—and this is the real departure from the *F. ananassa* varieties—recalls the aroma of the wild strawberry with its tantalizing sweet-sour taste. Spadeka's firm skin indicates it will freeze well. This is a variety for gourmet jams and marma-lades—again with a wild straw-berry flavor (cooking time is short to preserve the delicate flavor). Berries ripen around the same time but can stay on the plants up to a week after they begin to ripen. A single square yard of strawberry meadow can yield over four pounds of fruit.

Here are some brief instructions for their care: Before planting, pre-pare the soil by double digging (two spade-lengths' deep) and remove all weed roots. Recom-mended distancing: rows, 2 yards; distance in the row, 10 inches. Guide runners into the area be-tween rows to root.

Strawberries

Prevent weeds in your future meadow area by using a herbicide before planting.

After harvest, foliage may be mowed or allowed to stand. Only fertilize (again, after harvest) if the plants no longer attain 12 inches. When you no longer wish the meadow to expand, simply remove runners.

For many years, commercial growers have sought a machine to harvest. In a "mowing harvest," leaves and fruit stalks are mowed close to the ground. Leaves are then blown away and the stalks emptied of berries.

Another possible way to use Spadeka is as a ground cover. In a single season, it is possible to cover 3-square-yard areas. And even in shade, some harvest can be expected.

Many gardeners are looking for a strawberry that will produce in a small space without too much work and effort. Of all of the new hybrids so far, Spadeka fills the bill. Whether or not it will make its way into commercial plantings remains to be seen.

Blueberries

Blueberries are more often eaten than grown. Yet they are exceptionally handsome shrubs worthy of planting in the landscape.

Botany

There are a number of species of the genus *Vaccinium*—the whortleberries, *V. myrtillus,* the cowberries, *V. vitis-idaea,* the European cranberry, *V. oxycoccus,* and the bog bilberry, *V. uliginosum.* The blueberries that we know are derived from two American species, *V. corynbosum* and *V. australe.*

Blueberries are a large industry in this country—and one that is gaining popularity around the world. Bushes are robust, growing to 6 feet. In spring they are covered with white-to-pink, bell-shaped flowers. These are bisexual and self-fruitful. However, cross-pollination produces much higher yields and better berries. Several varieties should be planted where they can be pollinated by bees. After harvest, plants are a handsome addition to the garden with their attractive fall coloring.

Blueberry Berkeley.

European whortleberries (*left*). Blueberry Pemberton (*right*).

Blueberries

Varieties

The first blueberries were developed in America around 1906. They were selections from native species. Systematic crossing began in 1909. Since that time 50 to 60 varieties have been developed and introduced. Outside the United States, Dr. V. Heermann of Grethem, Germany, began to hybridize in the 1930s and produced a number of valuable species suitable for European conditions. Of these, the best is Blauweiss-Goldtraube from which the pure forms Ama and Heerma were derived. Of the American varieties—according to ripening time—the following are recommended: Earliblue, Bluetta, Collins, Meader, Patriot, Spartan, Ivanhoe, Blueray, Bluejay, Northland, Bluecrop, Bluehaven, Berkeley, Dixi, Herbert, Darrow, Coville, Lateblue, Elliott.

Ripening blueberries, Blauweiss-Goldtraube.

Blueberries

Where to Plant

For success with blueberries in the home garden, special pains need be taken to provide the right conditions for optimal growth.
Blueberries can withstand temperatures of 0 degrees F without any damage whatsoever. Their blooms are late enough not to be troubled by late frosts. They need a situation that is sunny, but out of the wind. Most important, they need an extremely acidic soil (pH 4.3–3.8) that is loose, deep, and rich in humus. Because most garden soils are only slightly acid to neutral, it takes a bit of doing to provide the heathlike conditions in which blueberries originate and thrive. Without additives, plants will sulk and die. But this doesn't mean that blueberries cannot be grown. There are two ways of growing them in an ordinary garden plot: The first is to dig a trench (5 feet × 5 feet × 2 feet); the second is to find containers, boxes, and large pots. Both trench and containers are filled with a mixture of acid peat, acid humus, sand and a little topsoil. Plants will last many years in either pots or trenches.

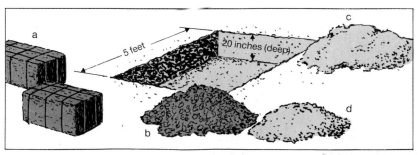

Trench for blueberries: (a) peat, (b) humus, (c) sand, (d) loam.

Blueberry grown in a pot of acid soil.

Blueberries

Planting

Plant blueberries in fall or very early spring. Space bushes in rows 9 to 10 feet apart and 5 to 6 feet apart within the rows. The best plants to buy are two- to three-year-old, well-rooted bushes from a nursery. These can be planted a little deeper than they were growing to encourage shoots from the base. Bushes should be firmly pressed into the ground and well watered. Cover the ground around the plants with peat, sawdust, or well-rotted manure.

Pruning

No pruning is necessary at planting time or for the first few years. Only when the plants are four to five years old will they need pruning to thin their dense growth and possibly to rejuvenate them. Both thinning and rejuvenating are handled as with currants: In winter, the oldest branches are cut back to the ground and new growth takes their places. No branch in the bush should be more than four to five years old.

Pruning to thin and rejuvenate a blueberry.

Blueberries

Soil

Blueberries are shallow-rooted, and the soil around them should never be mechanically worked. Best to mulch with about a 6-inch layer of straw, sawdust, peat, or laves. Weeds that get through the mulch should be cut away, and it is important to renew the mulch layer every two years.

Fertilizers

The best organic fertilizers are well-rotted manure, straw-manure, peat, and the like. The best chemical fertilizer includes trace elements but no chloride salts. Young plants need very little fertilizer. As a fast rule, 2 to 4 ounces per square yard are applied in early spring. If growth is weak or there are the yellow-green leaves that indicate hunger, an additional application of ammonium sulfate at the rate of 1 to 2 ounces per square yard may be administered in March to May. It is best to split this application in two. This is a good fertilizer for blueberries because it acidifies.

Diseases and Pests

Happily, blueberries are not much bothered by diseases or pests; the biggest problem is protecting the harvest from all kinds of birds. A single bush is easily covered with netting. Larger plantings have to rely on wind clappers, noisy scarecrows, flapping streamers, and recordings of the warning cries of crows or starlings. Once in a great while, an infestation of caterpillars, *Operophthera brumata,* occurs, but can be quickly controlled with insecticide. Branches stricken with the fungus *Godronia cassandrae* should be cut off and destroyed. A severe case may require one or two sprays in the fall, after the leaves fall, and a spring spraying with fungicide.

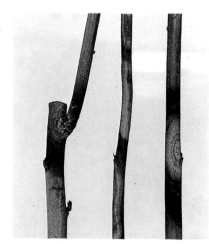

Blueberry stricken by the fungus *Godronia cassandrae.*

Blueberries

Harvest

Blueberries begin to bear full crops after the third or fourth year. A mature shrub yields between 5 and 12 pounds of berries—but much larger yields (over 25 pounds) are possible. Good care will keep a bush bearing for decades. Depending upon variety, fruits ripen from mid-July until September. Berries are completely ripe when they are fully blue and their aroma is strong.

One of the best things about blueberries is their ability to keep for long periods in the refrigerator. They transport well, too. The crop is a quick and easy harvest of precious berries.

The delicious berries are a quick and easy harvest.

Cowberries

The cowberry is one of the few fruits that is still gathered in the wild in Europe. As a fruit that is still in this somewhat primitive stage of development, it would seem that there would be few consumers and no market value. Just the opposite is the case: Cowberries are so versatile and have such an excellent taste that they are considered a delicacy.

The main problem is the difficulty of harvesting the berries in the woods, and only a portion of them do get harvested. Efforts have been made to cultivate them. At the beginning of the 1970s in Sweden, Finland, and Germany, the first efforts at cultivating cowberries were made. Although many questions remain to be answered about growing these plants, the work has yielded a small body of useful information that may be helpful to the gardener.

Introducing a new kind of fruit requires time and a pioneering

Selection of the first cowberry variety Koralle.

Cowberries

spirit. It has to be gone about slowly, patiently. But the home gardener can grow these berries with no constraints, enjoying the handsome plants and their delicious berries.

Botany

The cowberry, *Vaccinium vitis-idaea,* belongs to the family Ericaceae. With its creeping underground stems it grows into bushes 12 inches high. Bushy foliage and flower stalks grow from the creeping stems, and plants are evergreen. White, sometimes pink, bell-shaped flowers grow in multi-flowered clusters; the flower is bisexual and is pollinated by bumble bees, bees, and sometimes flies which cross-pollinate, although it appears that self-pollination is also possible. Fruit, too, hangs in thick clusters, turning from white to scarlet; the berries are shiny, round, about the size of whortleberries, and very seedy. Because they contain high amounts of citric acid, they are quite sour and for this reason, rarely eaten fresh. Instead they make excellent compotes, jams, wines, liqueurs, or hot wines.

In most cases plants bloom and fruit twice each year. The first flowering is in May or June, the second, July and August. The harvest from the first bloom is not important and is often neglected. Often plants—especially those growing on moors or heaths—are struck by a late-spring frost. Fruit from the second flowering is bigger and more uniform, and berries that don't ripen until fall or late fall are bigger and of better quality than summer berries.

Propagation

Cuttings are the only way to propagate uniform young plants. These are taken at the end of June until the end of August and grown under glass or plastic in a mixture (1:1) of sand and peat. By spring the new plants will have formed roots so that they can be transplanted into small (3-inch) pots. After another year or two, they may be planted in the garden.

Varieties

The first step toward creating a variety is selecting good specimens from the wild. In a Dutch arboretum an outstanding individual was found and christened Koralle. This is no clone, but a population of very similar plants which are particularly robust, exhibit regular growth and large berries.

Attractive fruit of the European cowberry Erntedank.

Three more true varieties were found in Germany growing on a northern moor: Erntedank, Erntekrone, and Erntesegen. Erntesegen is the most important selection, with its branches up to 16 inches tall and its light red, very large berries. This plant won the 1981 Gold Medal at the German Garden Show.

Where to Plant

Cowberries are found in many places in Europe—on heaths, moors, low mountains—even in the Alps up to 9,000 feet. Often they form the undergrowth in dry woodlands, especially under conifers, where the soil is sandy, stony, and rich in humus but infertile and acidic. In the garden, they require the same sandy, humus-rich, slightly acidic (pH 5–6) soil in full sun to semishade. For a garden without these attributes, additives like peat, woodland humus, and sawdust have to be worked into the soil to acidify it in the same manner as with blueberries. In areas where there is little snow, it is recommended to cover plantings with straw, evergreen boughs, or similar materials so that the evergreen leaves aren't burned in winter winds.

Cowberries

Planting and Care

The best time to plant is in fall or early spring. For the home gardener, bed planting is recommended. Five rows at a distance of 10 inches between rows and 10 inches between plants in the row are suitable. Because the plants send out many stems, the bed will soon grow into a dense ground cover.

Not much is known about the nutritional needs of the European cowberry. It doesn't need much. ½ ounce of a complete fertilizer with magnesium per square yard is plenty. Attempts to mulch with sand, sawdust, milled peat, and pine needles have proven advantageous. Keeping down weeds as well as preserving moisture make the maintenance of the bed easier. Not much is known about diseases or pests of these berries. Occasional reports about damage to plants in the wild by field mice, moles, insects, and fungal-disease carriers indicate that cultivated cranberries will need protection from these pests.

Harvest

While commercial growers may have machines for harvesting, the gardener has to rely on hand picking—at best with the aid of a picking comb. What sort of harvest can we expect? The Finns estimate that in their first experimental plantings the yield will reach 10 to 16 ounces per square yard; the Swedes predict harvests up to 2 pounds per square yard; a German grower reported yields up to 3 pounds per square yard. But only time will tell. In any case, growers are likely to find a good market for their fruit.

Cranberries

Although cranberries are familiar sights on holiday tables, they are infrequently grown in American gardens. They are commercially grown in northern areas where the climate, soil, and moisture suit their culture.

Botany

The cranberry, *Vaccinium macrocarpon,* belongs to the Ericaseae. The name comes from "crane's berry." A smaller variety is *V. oxycoccus.* They are low, evergreen, long-lived shrubs with long, thin, woody stems. These can grow to 4 feet or more during the growth period. Two or three years after planting, cranberries form a dense, closed mat or ground cover.

Vertical branches develop along the long runners on which flowers and fruit are carried. At the end of June to the beginning of July, after frosts are over, whitish pink, ¼-inch blossoms appear. At first they grow upright, but gradually bend over until they hang down. The bisexual flowers are protandrous and rely upon cross-pollination. The best pollinators are bumblebees and wild bees of which there are often insufficient numbers. For this reason, farmers often set beehives in and around the plantings. The production of large

From left to right: The cranberries *V. macrocarpon, V. oxycoccus,* and the cowberry *V. vitis-idaea.*

numbers of seeds is crucial for the development of large, well-formed fruits. About 75 to 100 days after bloom, either in September or October depending upon variety, the berries ripen. They are on the average nearly 1 inch long and over ½ inch wide, and may be round, oval, bell- or pear-shaped, red, dark red, or nearly blackish red.

Varieties

The number of cranberry varieties is unusually large. Most are selections from the wild. The newer ones are the products of systematic crosses. Some of the most common varieties are: Early Black, Howes, McFarlin, Searles, Ben Lear, Black Veil, and the especially large-berried Pilgrim.

Cranberries

Where to Plant

Cranberries prefer cool summers and endure winter temperatures of 0 degrees F and below without damage. Often a blanket of snow will accompany lower temperatures. Nevertheless, it is useful to cover the plants over the winter with straw, evergreen boughs, or similar material over long, cold periods when the soil is frozen and no water can be absorbed by the plants. One interesting practice is flooding the cranberries in mid-December and holding them in ice until the middle of March.

The cranberry is a bog plant. The best place for it is a moist moor or sandy soil, where—this is very important—the pH is quite acidic (4–5). Because most garden soil is slightly acid to neutral, amendments of peat, humus and other materials must be added. The plants need full sun.

Planting and Care

Plants are easy to propagate. 6-inch cuttings taken in spring as well as in July and August root quickly under glass in a mixture of peat and sand (3:1). The best time to plant is spring (April or May) or in fall (September to October). Set plants about 1 foot apart (ten to fifteen plants per square yard). After planting the most important chore until the planting has filled in is weed control. Water during droughts and give occasional fertilizer only if growth is poor, at the rate of ½ ounce per square yard. Too much nitrogen will foster the growth of foliage at the expense of fruits. Use naturally acidic preparations that are free of chloride salts.

Cranberries

Harvest

After four or five years, cranberries reach their peak. Because of their natural composition, which inhibits the growth of bacteria and fungi, the berries can be kept at temperatures of slightly above freezing for months on end. There are endless recipes for both cranberries and cowberries or mixtures of both. They are especially well suited as an accompaniment to game. But they are also excellent in juices, compotes, marmalade, honey, and the like. They are also used in puddings, ice creams, soups, candied fruits, cakes, cookies—even in soaps and perfumes, candles, aftershave, as well as other cosmetics. Cranberry plants make attractive garnishes and table decorations. There is a magazine called *Cranberry,* and even a Cranberry Queen!

Section of a cranberry stand of Early Black. Fruit is beginning to ripen.

Index

Index

Index